S0-BPN-308

PROFITABLE SOIL MANAGEMENT

PROFITABLE

PRENTICE-HALL, INC.
ENGLEWOOD CLIFFS, N. J.

SOIL MANAGEMENT

LEO L. KNUTI

Head, Department of Agricultural Education
Montana State College
Bozeman, Montana

MILTON KORPI

Vocational Agriculture Instructor
Dassel Public Schools
Dassel, Minnesota

J. C. HIDE

Professor of Soils
Montana State College
Bozeman, Montana

PRENTICE-HALL VOCATIONAL AGRICULTURE SERIES

PRENTICE-HALL VOCATIONAL AGRICULTURE SERIES

BEEF PRODUCTION. *Diggins and Bundy*

CROP PRODUCTION. *Delorit and Ahlgren*

DAIRY PRODUCTION. *Diggins and Bundy*

EXPLORING AGRICULTURE. *Evans and Donahue*

FRUIT GROWING. *Schneider and Scarborough*

JUDGING LIVESTOCK, DAIRY CATTLE, POULTRY, AND CROPS. *Youtz and Carlson*

LEADERSHIP TRAINING AND PARLIAMENTARY PROCEDURE FOR FFA. *Gray and Jackson*

LIVESTOCK AND POULTRY PRODUCTION. *Bundy and Diggins*

MODERN FARM BUILDINGS. *Ashby, Dodge, and Shedd*

MODERN FARM POWER. *Promersberger and Bishop*

POULTRY PRODUCTION. *Bundy and Diggins*

PROFITABLE FARM MANAGEMENT. *Hamilton and Bryant*

PROFITABLE FARM MARKETING. *Snowden and Donahoo*

PROFITABLE SOIL MANAGEMENT. *Knuti, Korpi, and Hide*

PROFITABLE SOUTHERN CROPS. *Walton and Holt*

THE RANGE AND PASTURE BOOK. *Donahue, Evans, and Jones*

RECORDS FOR FARM MANAGEMENT. *Hopkins and Turner*

SHEEP PRODUCTION. *Diggins and Bundy*

SOIL — USE AND IMPROVEMENT. *Stallings*

SOILS — AN INTRODUCTION TO SOILS AND PLANT GROWTH. *Donahue*

SWINE PRODUCTION. *Bundy and Diggins*

USING ELECTRICITY ON THE FARM. *Hamilton*

YOUR FUTURE IN POULTRY FARMING. *Goodman and Tudor*

Library of Congress Catalog Card No. 62-7584.

Printed in the United States of America 72944-E

ABOUT THIS BOOK

Illustrated with more than 225 carefully selected photographs and specially prepared diagrams, *Profitable Soil Management* contains a full treatment in nontechnical language of soil science, soil management, and soil and water conservation.

Contents include (1) an introductory explanation of what soil management is and why it is important; (2) a scientific definition of soil and detailed accounts of its origin, classification, and major parts; (3) an analysis of how plants grow and the effect of environment upon growth; (4) a complete treatment of soil improvement, with sections on essential soil elements, soil reaction, acid soils, alkali soils, fertility tests, commercial fertilizers, manures, drainage, and irrigation; (5) a description of conservation of soil and water in the nation, on the individual farm, and within the cultivated field; (6) a treatment of tillage practices and equipment; and (7) a complete chapter on land judging.

There are a glossary and index.

For valuable suggestions, research materials, and illustrations, the authors are indebted to state agricultural colleges and their personnel, high school vocational agriculture teachers, county agents, practicing farmers, commercial firms, public and private agricultural publications, and the United States Department of Agriculture with its research facilities and especially its Soil Conservation Service. Extensive help was provided by the personnel of Montana State College.

Reviewers included Professor Richard H. Wilson, Assistant Professor, Department of Agricultural Education, The Ohio State University, Columbus, Ohio; Mr. G. R. Cochran, State Supervisor, Agricultural Education, Vocational Division, St. Paul, Minnesota; and Mr. Elvin Downs, Assistant Supervisor, Department of Agricultural Education, Salt Lake City, Utah.

CREDITS FOR ILLUSTRATIONS

Courtesy of Allis-Chalmers, 1, 3, 27, 37, 48, 63, 64, 91, 92, 107, 108, 142, 159, 183, 205, 245, 252, 261, 274, 277, 279, 288, 297, 305, 306, 312, 315, 317, 318, 329, 333. Courtesy of the American Potash Institute, 36, 171. Courtesy of Russell S. Anderson, Agricultural Agent, Hartford, Connecticut, 164, 165, 166, 168. Courtesy of the University of Arkansas Agricultural Experiment Station, 293. Courtesy of Armor Metal Products, Helena, Montana, 248. Courtesy of Beckman Instruments, Inc., 134, 162. Courtesy of Wray Birchfield, USDA, 58 (F. 4-14). Courtesy of B. A. Brown, Agronomy Department, Storrs (Connecticut) Agricultural Experiment Station, and *Hoard's Dairyman,* 189, 193. Courtesy of Buckner Manufacturing Company, Inc., 228. Colonial Williamsburg Photograph, 253. Courtesy of the University of California, 71 (right). Courtesy of Rex Campbell, Montana State College, 12, 338, 339 (F. 21-13), 340. Courtesy of John Deere, Moline, Illinois, 40, 184, 214, 226, 232, 257, 319, 320. Courtesy of Dr. L. D. Doneen, 154. Courtesy of H. M. Ellis, North Carolina Agricultural Extension Service, and *Hoard's Dairyman,* 230. Courtesy of *Farm Journal,* 255. Courtesy of *Farm Management* and Les White, 170. Photo by the Greater North Dakota Association, 5 (left), 247, 264, 313. Courtesy of Grumman Aircraft, 2 (F. 1-1). Courtesy of *Hoard's Dairyman,* 123, 131, 198, 230. Courtesy of Illinois Natural History Society, 58 (F. 4-13). Courtesy of the University of Illinois, 327. Courtesy of International Harvester, *viii,* 39, 140, 143, 195, 200, 201, 212, 285, 309, 310, 334. Courtesy James Madison Junior High School, Los Angeles, California, 111. Courtesy of Paul Jesswein, Montana State College Information Department, 43, 46 (F. 4-5), 220. Courtesy of Edward J. Karlin, 57. Courtesy of Dr. Harry Keener, University of New Hampshire, 120. Reprinted from "Climate and Soil," by C. E. Kellogg, 1941 Yearbook of Agriculture: *Climate and Man,* 24-25. Courtesy of Les Kinzell, Montana State College Information Department, 18, 19, 22, 34, 79, 84, 89, 213, 216, 294, 300, 331. Courtesy of A. L. Lang, University of Illinois, 80. Courtesy of T. M. McCalla, Bacteriologist, USDA Agricultural Research Service, 112. Courtesy of Gerald McKay, Minnesota Agricultural Extension Service, 126. Courtesy of Robert Meyer, Reedsburg High School, Reedsburg, Wisconsin, 174. Courtesy of Michigan State University, 71 (left). Courtesy of the University of Minnesota, 87. Courtesy of Mississippi Extension Service, 180. Courtesy of Missouri Farm News Service, 139. Courtesy of Montana State College Horticulture Department, 95. Courtesy of Montana State College Information Department, 5 (right), 124. Courtesy of E. A. Naphan, SCS, Reno, Nevada, 16, 240. Courtesy of the University of Nevada, 148, 154. Courtesy of Agronomy Department, University of New Hampshire, 117. Courtesy of the New Holland Machine Company, 202. Courtesy of the Nitragin Company, Inc., 44 (F. 4-1), 49. Courtesy of North Carolina Department of Agriculture, 160. Courtesy of North Dakota State University, 20. Courtesy of Northwest Rain Bird Sales, 6. Courtesy of Extension Division, Oklahoma State University, 342-343. Courtesy of Karl Parker, Montana State College, 106. Courtesy of Pennsylvania Agricultural Extension Service, 4, 11, 144. Courtesy of Percival Refrigeration and Manufacturing Company, 109. Courtesy of *Plant Food Review,* a publication of the National Plant Food Institute, 55. Reprinted from *Plant Food Review,* a publication of the National Plant Food Institute, 187. Courtesy of Chas. Pfizer & Co., Inc., 52. Courtesy of Choe Noel, Hydroponics, Inc., of Montana, 121. Courtesy of E. R. Purvis, Rutgers University, 116. Courtesy of Paul Ratliff, Baldwin, Mississippi, 83, 101. Courtesy of Bureau of Reclamation, 2 (F. 1-2), 15, 85, 151, 256, 258. Courtesy of Rothamsted Experimental Station, Harpenden, Herts., England, 177, 178. Courtesy of Fred Sanford, Montana State College Information Department, 44 (F. 4-2), 46 (F. 4-4), 51, 65, 66, 67, 77. Courtesy of George Sharpe, Agricultural Extension Service, West Virginia University, 129, 324, 337 (F. 21-9, F. 21-10), 339 (F. 21-12, F. 21-13). Courtesy of "Shur-Rane" Irrigation, 227. Courtesy of the Soil Conservation Service, 7, 29, 33, 35, 72, 73 (F. 5-8, F. 5-9), 74, 76, 108, 118, 150, 161, 191, 206, 207, 208, 210, 218, 223, 225, 237, 241, 250, 262, 266, 267, 268, 273, 282, 287, 299, 302, 304, 323, 336, (F. 21-7, F. 21-8). Sunkist Photo, 103. Courtesy of Texas Agricultural Extension Service, 113, 269. Courtesy of W. R. Thompson, Mississippi State College, 271. Courtesy of M. K. Thornton, Texas Agricultural Extension Service, 60. Courtesy of Emil Truog, Professor Emeritus, University of Wisconsin; reproduced by permission from August 10, 1955 issue of *Hoard's Dairyman,* Copyright 1955 by W. D. Hoard and Sons Company, Fort Atkinson, Wisconsin, 131. Courtesy of Emil Truog, Professor Emeritus, University of Wisconsin, and *Plant Food Review,* a publication of the National Plant Food Institute, 55, 137. Courtesy of Union Pacific Railroad, 13, 69, 127, 234, 235. Courtesy of United States Department of Agriculture, 99, 100, 104, 105, 147, 181, 196, 198, 238, 304, 316. Courtesy of the U. S. Navy and SCS, 299. Courtesy of Vermeer Mfg. Co., 54. Courtesy of Les White and *Farm Management,* 170. Courtesy of University of Wisconsin, 155.

CONTENTS

The Importance of Our Soil and Its Management

Soil is the mineral and organic matter that supports plant growth on the earth's surface. It is a mixture of particles of rock, organic materials, living forms, air, and water.

Soil is the most important natural resource of America. Each person in our nation has a vital interest in seeing that our soil is improved rather than destroyed. Soil concerns farmers, communities, counties, states, and the nation.

Soil Serves Us All

The person with the greatest concern for the soil is the farmer. His standard of living is dependent upon its productivity. Future farmers, who expect to make their living on the farm, must gain information on how to use land resources efficiently. Good land is reflected in prosperous-looking farmsteads, healthy livestock, and good crops.

It is important to remember, however, that it is not only the farmer who benefits or suffers from how the soil is used. Our communities, counties, states, and nation are dependent upon the fruits of the soil for food, clothing, building materials, and basic elements of thousands of manufactured products. Two and one-half acres of farmland are required to feed each American. Soon we will be so numerous that there will be more Americans than farmlands to support them at present productivity rates. Upon the continuing improvement of productivity thus depends

Figure 1-1: Fighters and Farms. The taxes paid by the farmers whose fields spread below help keep our America strong.

the future well-being of our country. Even today, the prosperity of the farmer determines the incomes of many of our merchants and industries, and contributes to the ability of our governments on every level to provide services and security to us all through our taxes. Our many land-grant colleges and universities were founded and continue to be partially supported by the income from farmlands given to the states by the Federal government

Figure 1-2: The Grand Coulee Dam. America recognizes the value of its farmers by undertaking vast reclamation projects.

Figure 1-3: The American Harvest. Our bounty is one of the wonders of our time.

for this purpose. Hundreds of counties all across America are paid substantial percentages of the rentals derived from Federal lands used for grazing and other farming purposes. Such counties spend this money for education and other public purposes.

Discontent and unrest are often associated with low standards of living. Our nation has adopted a foreign policy of aid to the underdeveloped countries. In large measure this aid has been designed to increase soil productivity through American know-how. Our surplus food has been used to provide temporary relief to these less fortunate countries.

It is in recognition of such values that thousands of American communities support vocational agriculture courses in their high schools. Assistance to farm operators is usually provided at the county level through the county agent, Soil Conservation Service personnel, and vocational agriculture instructors. Each of our states has at least one agricultural college. State agricultural experiment stations undertake research to increase farm efficiency. Also at the state level are the administrative offices for the Extension Service, vocational agriculture, and the Soil Conservation Service. States have passed laws which establish soil conservation districts. The Federal government sponsors

Figure 1-4: The Well-managed Farm. Woodland, pasture, and contour strips lead to handsome barns and a fine home.

agricultural research services, the Soil Conservation Service, the Forest Service, and the Bureau of Reclamation.

Since soil is the basic raw material of farming, these vast public efforts in education, assistance, and research might be said to be primarily concerned with just one thing, the profitable management of our soils.

The Importance of Soil Management

Soil management is the science of the origin, nature, use, protection, and improvement of soil. It includes the definition and classification of soil, and detailed considerations of its minerals, organic matter, plant and animal life, and moisture. The means by which plants grow are of importance to it, as are the hereditary and environmental factors controlling that growth, and the chemical elements needed. Finally, it involves the means by which soil is adapted to growth requirements. This book considers many aspects of soil management, such as acidity and alkalinity, liming, fertility tests, commercial (chemical) fertilizers, manures and other organic fertilizers, drainage, irrigation, conservation on the national scale, conservation on the farm level, conservation within the cultivated field, tillage practices and equipment, and land judging.

Not until as late as 30 years ago did most Americans realize that in large areas of our country up to one half of the topsoil has been removed by erosion. Equally significant has been the removal of large quantities of plant nutrients from the soil through leaching and the harvesting of crops and livestock. There have also been great losses of organic matter, not only decreasing fertility but making the soil more difficult to till.

The Agricultural Revolution. This new awareness greatly increased the pace of the revolution in agriculture which began a little more than a century ago, with the invention of the mechanical reaper and the founding of our first agricultural colleges. For the first time, the scientific method was systematically applied to all aspects of farming, not only by professional researchers, but by many progressive farmers. Even more important, almost all farmers began to study and apply the results of experiments which already had been completed, such as the Morrow Plots experiments of the University of Illinois, demonstrating the value of crop rotation.

Soil Management Applied to Farming. Today modern scientific instruments and chemical analysis can usually determine the essential food and other conditions lacking in soils. Earlier, attention to soil was centered on such major fertility elements as

Figure 1-5: Science, Progress, Prosperity. Agricultural experiment stations such as the one at Dickinson, N.D. (left), and agricultural colleges such as Montana State (right) raise our efficiency.

Figure 1-6: Water and Windbreak. The newest equipment is often the most profitable equipment. Here, sprinkler irrigation is used with a windbreak, which cuts evaporation losses and keeps water-application even.

phosphorus, nitrogen, and potassium. Now it is known that lack of minor, or *trace,* elements, such as boron, zinc, and iron, may be just as important, and that such other factors as soil texture and water storage capacity cannot be overlooked. Under certain conditions, deficiencies in any of these result not only in poor, expensively produced yields but in plant and livestock products deficient in elements necessary for human health.

Soil Management and Crop Production. With the help of our knowledge, every farmer can immediately increase his crop yields and his income by taking advantage of management practices such as liming, correction of alkali conditions, mineral and organic fertilization, drainage, irrigation, conservation, and land judging. How much room there remains for improvement is evidenced by the facts that, under the most favorable conditions, corn yields of over 200 bushels per acre and wheat yields of over 100 bushels per acre are possible, but that the national average yield per acre for corn is about 50 bushels and for wheat about 20 bushels. In addition to better yields at lower cost per bushel, many farmers will find that rich soil produces healthy plants which are freer from diseases and less likely to be attacked by insects.

Soil Management and Livestock Production. Farmers know, too, that the food that livestock eat is reflected in their health, and that the healthiest animals are seen on farms and ranches that have productive lands. Even in regions where other soil conditions are good enough to produce feed, farmers have observed livestock chewing on boards and bones, attempting to correct mineral deficiencies in their diets. Certain food elements are essential for proper growth, reproduction, and production, and a lack of any of these causes stunted animals, less milk, fewer eggs, less wool, and smaller litters.

Deficiencies of various plant food elements also are responsible for specific livestock ailments. Lack of calcium, for instance, causes weakened bones, lower milk production, and rickets in young animals. Lack of phosphorus lowers breeding ability and is another cause of rickets in young animals. Lack of iodine produces an enlarged neck or goiter in calves, colts, and lambs, and hairlessness in pigs. Lack of cobalt causes a loss of appetite in cattle. Lack of iron and copper produces pale blood and skin color in cattle, sheep, and goats, and "thumps" (anemia) in pigs. A manganese deficiency causes slipped tendons and lower hatching rates in poultry, and lameness in pigs.

All of these deficiency diseases today can be largely prevented by modern soil management.

Rewards and Danger

As a result of the application of a century of research, farming is so altered that many farmers themselves are not fully aware of how complete the change has been. Plowing, sowing, cultivation, and harvest, which used to require days of backbreaking labor, can now be accomplished in a few hours by machines. Plant diseases and insect infestations which once left thousands of square miles of bare fields in their wakes can be controlled in minutes by chemical dusts and sprays. Some new chemicals will weed for the farmer, others being developed may help

Figure 1-7: An American Home. The envy of many a city-dweller, this well-landscaped country home in Montana has the added attraction of having been an historic roadhouse for the Vigilante Stagecoach Line.

improve the structure of soil. His crop yields can be doubled in a few hours by the application of lime or fertilizer. He can determine to the merest trace what nutrients his crops require. Wherever he lives, he can choose crops and livestock suited to his climate. He can farm the desert and the swamp if he wishes. He no longer need helplessly watch while his farm and future are blown away by winds or washed away by rains.

By these and other means, the standard of living of our best farmers has become one of the wonders of our time. Russian visitors have been incredulous at the fine houses, the cars, and the television aerials lining the highways for all to see. Some farmers own planes; others spend vacations in Florida and see nothing remarkable in trips to Europe. It is no longer necessary, in the words of the old joke, for the farm boy to go to the city to become rich enough to move back to the farm. All the advantages of the best suburban living may be his without leaving home.

At the same time, the efficiency of agricultural science has made it ever more difficult for the inefficient farm to survive. The best methods have lowered production cost per bushel to the point that to farm poorly has become fantastically expensive. A farmer's living expenses, his real estate taxes, and the cost of plowing, seeding, cultivation, and harvest remain much the same whether the result of his labor is thousands of high-quality bushels for the market, or a few low-quality hundreds. Not to

Figure 1-8: The Farmer's Choice. The corn and wheat displayed at the left of the sign marking the famous Morrow Plots show the results of careful soil management; the corn and wheat displayed at the right show the results of its long neglect.

investigate the newest machinery, not to lime, not to fertilize, not to drain, not to irrigate, not to use green manures and rotations and contour plowing and all the other techniques of soil management—in short, not to keep up with the latest developments—these are extravagances which many farmers have found ruinous.

Summary

Soil is the mineral and organic matter that supports plant growth on the earth's surface. It is a mixture of particles of rock, organic materials, living forms, air, and water.

Soil is the most important natural resource of America. It furnishes us with food, clothing, building materials, basic elements of thousands of manufactured products, markets for stores and industries, taxes for public services, support for land-grant colleges and universities, rentals from public lands, and food and information useful in the conduct of our foreign policy.

In recognition of the value of soil, services are provided to farmers and future farmers by our communities, counties, states, and Federal government.

Soil management is the science of the origin, nature, use, protection, and improvement of soil.

Not until as late as 30 years ago did most Americans realize the need for soil management. This realization increased the pace of the agricultural revolution, in which the scientific method is applied to farming. Today, farmers who use soil management can greatly increase crop and livestock yields and their own standards of living. Farmers who do not use modern soil management find it ever more difficult to survive.

Study Questions

1. What is soil?
2. What are nine reasons why soil is important to all Americans?
3. How is soil productivity encouraged by communities, counties, states, and the Federal government?

4. What is soil management?
5. What was the "agricultural revolution"?
6. What are the yields, in bushels per acre, of corn and wheat raised under the best present conditions, and what are the average yields per acre for these crops?
7. In what ways has the application of soil management and other agricultural sciences benefited the progressive farmer's standard of living?
8. Why is it more expensive to farm without soil management than with it?

Class Activities

1. Invite the head of your local Chamber of Commerce to class to tell you how farmers are important to the prosperity of the community. Or conduct a survey to discover what the purchases, contributions, and taxes of farmers mean to local stores, churches and other charities, and public services. Be sure to send your figures to the local newspaper.
2. Invite your county agent to class to describe the government services locally available to farmers.
3. Find pictures to illustrate farming methods and farmhouses of the following: (1) the ancient Egyptians; (2) the ancient Greeks; (3) the ancient Romans; (4) Europeans during the Middle Ages; (5) American pioneers; (6) Americans of the late nineteenth century; (7) Americans of the early twentieth century; (8) Americans during the 1930's; (9) Americans today.
4. Invite your science teacher to class to describe the *scientific method.*
5. Find pictures and articles in magazines, newspapers, and books published during the 1930's to discover what happened in America to draw public attention to the need for soil management. Bring them to class to show to the other students.
6. Secure local production figures for your most common crops. Get these figures, and compare them with the national averages of production of those crops, and with prize yields of those crops. Save the figures, and when you complete this course use them to determine (1) exactly what must be done on a specific local farm to match the best yields, (2) how much it might cost to do this, and (3) whether the owner of the farm would make a profit on *each* of these improvements at current prices.

What Is Soil?

Soil, you remember, *is the mineral and organic material that supports plant growth on the earth's surface. It is a mixture of particles of rock, organic materials, living forms, air, and water.*

In this chapter you will learn how soil is created, what kinds of rock are used to form it, and how it is classified. In the three chapters following you will learn about the organic materials, living forms, air, and water in the soil. Soils in different parts of the world are similar when produced from similar parent materials and under similar climatic and physical conditions.

Origin of Soil

Soil only appears to be permanent and changeless. Actually, something is always being added or taken away. The changes tend to balance each other, and therefore soil appears to be stable. Soil materials include the remains of everything that has lived on the earth, in various states of decay and disintegration.

Soil starts out as rock. There are many different kinds of rock, from the hardest flint to soft rotted stone that turns to powder in your hands. But these rocks can all be divided into three main groups according to how they were formed. One kind was formed from material melted in the terrifically hot furnaces of volcanoes. Another kind was formed from material laid down by rivers, by moving glaciers, and by wind. A third kind was formed by the squeezing action of tremendous masses of earth,

Figure 2-1: The Birth of the Land. Can you identify the forces that are slowly turning these Western badlands into soil? What forces might account for the rounded tops of the rocky hills, for instance?

or by the intense heat generated by this pressure. Some rocks are rich in the chemicals needed by growing plants.

Rocks are generally broken up by powerful forces. For instance, a rock is split apart by expansion and contraction as a result of heat and cold. Water freezes and expands in cracks, wedging a rock apart as you would split a tree trunk with wedges and a mallet. Rivers and glaciers and landslides and avalanches cause small fragments of broken rock to rub against each other until they are ground up like wheat under a mill wheel. Winds sweep sharp grit across rocks, scouring them like some giant with a huge piece of sandpaper. Chemicals dissolved in soil water slowly eat rocks away. The combination of these processes is called *weathering*. This weathering of rock has been going on for ages and is still going on.

Next in the process of soil formation, a tiny plant begins to grow, perhaps a lichen or a tiny moss. When the plant dies, its substances decompose and combine with ground-up rock material. Gradually more plants take hold. Plants with deep roots draw materials from below the surface, and when the plants die, these materials are added to the surface material. Swarms of microorganisms become active, breaking down the remains of dead plants and turning them into humus. As this process continues, there is no longer just ground-up rock.

Soil microorganisms seek food not only from organic materials but also from the rock fragments which are combined with organic materials. Materials produced during decay include some acids which further work on rock fragments, modifying them. Thus the end product is a complex mixture of new and original minerals and partially decomposed organic materials. This mixture is what we call *soil.*

Soil Orders. There are five factors, then, that are responsible for soil: (1) parent material; (2) climate; (3) variations in the earth's surface, such as hills and valleys, which aid or retard the action of climate; (4) plant and animal life; and (5) time. Depending on how these factors interact, soils are divided into three great *orders.*

Zonal or "normal" soils are those that have developed fully, according to the ordinary conditions of the regions where they are found.

Intrazonal soils also have developed fully, but they have special characteristics because of local, rather than regional, conditions. Bog soils and alkali or saline soils, which develop in low places, are examples of intrazonal soils.

Azonal soils are those that have not developed fully. These include shallow soils on steep slopes, where soil cannot accumulate, and recently deposited sediments of streams. Azonal soils lack *horizons,* which are explained later in this chapter.

Materials and Processes of Soil Development

Now that you understand the formation of soil in a general way, we can apply some of the technical terms assigned to the materials and processes of soil development.

Figure 2-2: Mt. Hood. This beautiful peak, towering above the orchards of Oregon, is an extinct volcano. Do you know what kind of rock is formed by the hardening of molten materials from within the earth?

Materials of Soil Development. Soils are formed from (1) rocks, rock fragments, and minerals, and (2) organic materials.

Rock. Rocks from which soil materials are formed are classed as (1) igneous, (2) sedimentary, and (3) metamorphic.

Igneous rocks have been formed by the hardening of molten materials that originated within the earth. Kinds of igneous rocks include granite, syenite (chiefly feldspar), basalt (dark gray to black), and rhyolite, a lava form of granite.

Sedimentary rocks include those rocks formed from the consolidation of particles laid down in early geological ages. Examples are limestone, sandstone, siltstone, shale, and conglomerate.

Metamorphic rocks are those that have resulted from profound alterations of igneous and sedimentary rocks through heat and pressure. Examples include slate, marble, quartzite, gneiss, and schist.

Not until rock has been partially broken down is it called *parent material.*

Organic Parent Material. Two classes of organic parent materials are (1) peat and (2) muck. Peat deposits are formed in moist places where organic matter forms more rapidly than it decomposes.

Peat differs from muck in that the organic remains in peat are sufficiently fresh and intact to permit identification of plant forms. Peat is turned to muck when the organic materials have been decomposed to the extent that plant parts cannot be identified. Muck has a higher mineral or ash content than peat.

Transported Soil Parent Material. Much of the soil parent material of the United States has been moved from its place of origin. Transported soil materials are usually named according to the force responsible for their movement. Forces which transport soils are (1) water, (2) wind, (3) gravity, and (4) ice and glaciers.

Water-Deposited Soil Parent Material. Soil parent materials transported and deposited by water are named (1) *alluvium,* (2) *lacustrine deposits,* (3) *marine sediments,* and (4) *beach deposits.*

The most important water-deposited soil material is alluvium. Alluvium consists of soil sediments deposited by streams. It may

Figure 2-3: The Columbia Basin. This deep, fertile soil has been transported hundreds of miles from the places where it is thought to have originated.

occur in terraces or "bench lands" well above present streams, or in the normally flooded bottom lands of existing streams.

Lacustrine deposits consist of soil parent materials that have settled out of the quiet waters of lakes.

Marine sediments are soil parent materials which have been deposited by the waters of oceans and seas.

Beach deposits mark the shore lines of old seas or lakes. They include beaches of glacial lakes such as Lake Agassiz, now the Red River Valley of North Dakota and Minnesota. These deposits are often gravelly, cobbly, and stony.

Wind-Deposited Soil Parent Material. Wind-deposited soil material includes (1) sand, (2) volcanic ash, and (3) silt-like material called *loess.*

Sand deposits include the dunes on the leeward side of large bodies of water and on sandy deserts.

Volcanic ash deposits are found in Kansas, Nebraska, and Montana. Such soils are very porous and light in weight.

Loess was deposited in the central United States. This material was derived in part from sediments of huge rivers which were fed by continental glaciers, probably in the Rocky Mountains.

Gravity-Deposited Soil Parent Material. Colluvium soil parent materials are those moved primarily under the influence of gravity. They are to be found at the bases of sharp slopes. Forces of movement aiding gravity are (1) frost action, (2) soil creep, and (3) local wash.

Glacially-Deposited Soil Parent Material. The soil materials deposited directly by ice are commonly referred to as *glacial till.* Till is a mixture of clay, silt, sand, gravel, and sometimes boulders. Till may be found in *ground moraines,* which are long low hills. Fairly level glacial moraines also occur. In addition, glaciers are responsible for *outwash plains*—flatlands formed beyond the moraines by streams from the melting ice. Glacial lake deposits range from coarse gravel, notably along ancient beach lines, to fine clays.

The Soil Profile

If you dig very deeply into the soil, in making a trench, one of the things that is noticeable is that the ground lies in distinct layers. These layers are called *horizons,* and are classified according to the amount of weathering and decay that has taken place.

"A" Horizon. The top, most "finished" layer, the *"A" horizon,* may be from a few inches to a foot or more deep. It is a different color, usually darker, than the lower layers. The "A" horizon is where life, in the form of plant roots, bacteria, fungi, and insects, is most abundant in the soil. This horizon, roughly corresponding to the topsoil, is the layer usually cultivated for crops.

"B" Horizon. Below the "A" horizon is the *"B" horizon,* commonly called the "subsoil." This "B" horizon is often finer in texture than the "A" horizon.

Both horizons combine to make what soil scientists call the "true soil" or *solum.* Physical, chemical, and biological forces have been working on the two horizons long enough to transform material in them from what it once was to a new

Figure 2-4: The Soil Profile. Identify the "A" horizon, the "B" horizon, and the "C" horizon. Note the relatively light, leached topsoil and the accumulation of clay in the subsoil.

material—soil. These changes are what make the "A" and "B" horizons true soil. The average composition of the soil solum is: 45 per cent mineral, 5 per cent organic matter, 30 per cent water, and 20 per cent air.

"C" Horizon. Below the true soil is a third layer, the *"C" horizon,* which has not yet been turned into true soil. This horizon is called the *parent material* of the soil. As you may remember, parent material is material that has undergone some change from the original rock, but not enough to make it true soil.

"A," "B," and "C" are the three principal horizons, but in many soils each layer can be divided into secondary layers called "A^0," "A^1," "A^2," "B^1," "B^2," and so on. The three horizons with their individual parts constitute the *soil profile.*

Factors Used to Describe Soil

The major factors used in differentiating among different soils and soil horizons are (1) color, (2) texture, and (3) structure.

Soil Color

The main coloring material in surface soils is produced by organic matter. The dark color of topsoil usually indicates presence of organic matter.

In subsoils, color is usually due to the iron present and the form in which it exists. The form in which iron exists depends on the degree of aeration of the soil. If iron exists in the ferrous (FeO) form, the color of the subsoil is gray; if it exists in the ferric (Fe_2O_3) form the color is red; and if it exists in the limonite ($2Fe_2O_3 3H_2O$) form, the color is yellow. Silicate minerals usually tend to be white in color. Thus the intermediate colors of soil usually result from various combinations of white, black, gray, red, and yellow.

Soil Texture

A close examination of the soil shows that it is not a solid mass. It consists of countless numbers of particles of different kinds and sizes. Some of the particles are fragments of rock, which

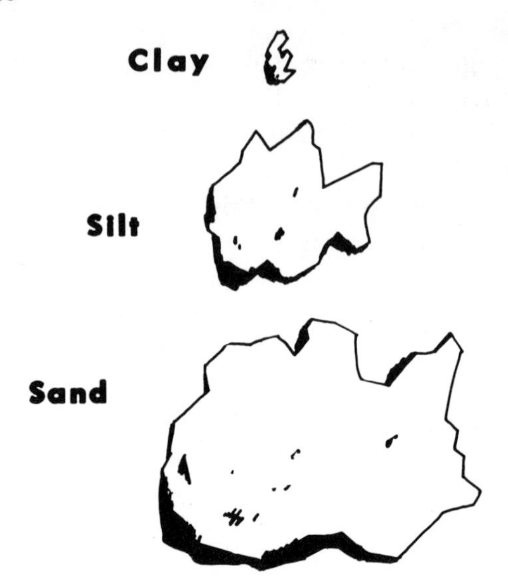

Figure 2-5: Relative Sizes of the Three Major Soil Particles. The proportions of sand, silt, and clay particles—here much magnified—in a soil determine its texture.

decrease in size from gravel to sand to the finest clay. Some are bits of decayed organic matter. Different soils have different proportions of coarse and fine particles, and this is what gives each soil its characteristic texture or feel. Some soils are coarse and gritty, some are smooth and silky.

Soil texture, then, is the property of the soil controlled by the size of the individual soil grains or particles, or the fineness or coarseness of the units of the soil mass. Soil is usually made up of particles of widely varying size. Soil textural terms express the average or combined effect of all these grain sizes.

In the laboratory, texture is determined by a mechanical process of separating the soil into groups of grain sizes. The three major size particles in soils, from largest to smallest, are sand, silt, and clay.

In the field, texture is determined by the "feel" of the moist soil mass when rubbed between the fingers. The following statements describe the physical characteristics of the basic textural groups.

Sand. Sand, as a soil texture type, is loose and single-grained. The individual grains can readily be seen or felt and have little tendency to stick together. Squeezed in the hand when dry, sand will fall apart when the pressure is released. Squeezed when moist, sand will form a cast, but will crumble when touched.

Sandy Loam. A sandy loam is a soil containing much sand, but with enough silt and clay to make it somewhat coherent. The individual sand grains can readily be seen and felt. Squeezed when fairly dry, sandy loam will form a cast which will readily fall apart, but if squeezed when moist, a cast can be formed that will bear careful handling without breaking.

Sands and sandy loams are classed as coarse, medium, fine, or very fine, depending upon the proportion of the different-sized particles that are present.

Loam. A loam is a soil having a relatively equal mixture of the different grades of sand, silt, and clay. It is mellow with a somewhat gritty feel, yet fairly smooth and slightly plastic. Squeezed when fairly dry, it will form a cast that will bear careful handling. A cast formed by squeezing moist loam can be handled quite freely without breaking.

Silt Loam. A silt loam is a soil having a moderate amount of the fine grades of sand and only a small amount of clay, with over half the particles being of the medium size, called *silt*. When dry, it may appear quite cloddy, but the lumps can be readily broken, and when pulverized it feels soft and floury. When wet, the soil readily runs together and puddles. Either

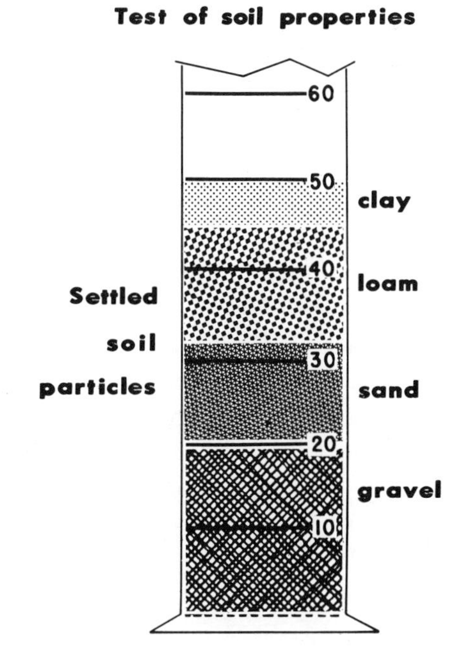

Figure 2-6: Separated Soil. You can determine the texture of a soil by adding a representative sample to water, shaking vigorously until the soil is in suspension, and then allowing the mixture to settle. The largest, heaviest particles will settle at the bottom, the lightest and smallest at the top.

Figure 2-7: Very Fine Sandy Loam. Use the chart of soil textural classes to give the first and second groupings for this soil. Can you find the division between the "A" and "B" horizons?

dry or moist, it will form casts that can be freely handled without breaking. If squeezed between the thumb and finger, it will not "ribbon," but will have a broken appearance.

Clay Loam. A clay loam is a fine-textured soil which breaks into clods or lumps that are hard when dry. When the moist soil is pinched between the thumb and finger, it will form a thin "ribbon" which will break readily, barely sustaining its own weight. Moist clay loam soil is plastic and will form a cast that will bear much handling. When kneaded in the hand, it does not crumble readily, but tends to work into a heavy compact mass.

Clay. Clay is a fine-textured soil that forms very hard lumps or clods when dry, and is quite plastic and usually sticky when wet. When moist clay soil is pinched between the thumb and fingers, it will form a long, flexible "ribbon."

Gravelly or Stony Soils. All of the above classes of soil, if mixed with a considerable amount of gravel or stone, may be classed as gravelly sandy loam, gravelly clay, stony sandy loam, stony clay, and so on. In general, stones must be larger than four inches in diameter to be called "stones," and less than four inches in diameter to be called "gravel." In order for a soil to be considered stony or gravelly, there must be enough stone or gravel present in the surface soil to definitely interfere with cultural practices. Approximately 20 cubic yards of rock or gravel per acre is necessary for a soil to be called stony or gravelly.

General Grouping of Soil Textural Classes
(USDA Classification)

First Grouping	*Second Grouping*	*Basic Soil Textural Class Names*
(I) *Sandy soil*	(1) Coarse-textured soil	Sand Loamy sand
(II) *Loamy soil*	(2) Moderately coarse-textured soil	Sandy loam Fine sandy loam
	(3) Medium-textured soil	Very fine sandy loam Loam Silt loam Silt
	(4) Moderately fine-textured soil	Clay loam Sandy clay loam Silty clay loam
(III) *Clayey soil*	(5) Fine-textured soil	Sandy clay Silty clay Clay

The above classification permits classifying soils first into three general groups as (I) sandy soil, (II) loamy soil, and (III) clayey soil. The second grouping of five general classes is (1) coarse-textured soil, (2) moderately coarse-textured soil, (3) medium-textured soil, (4) moderately fine-textured soil, and (5) fine-textured soil. The basic soil textural classes fall in the further general groupings shown above.

Soil Structure

Terms used to describe the different horizons of a soil profile include *structure* as well as *color* and *texture*. *Texture*, you remember, has reference to the size of soil particles. *Structure* refers to the *grouping* or *arrangement* of soil particles. *Soil aggregate* is a term used in describing a single mass or cluster of soil particles, such as a clod, crumb, or granule. *Soil tilth* refers to the fitness of a soil for cultivation, as determined by the

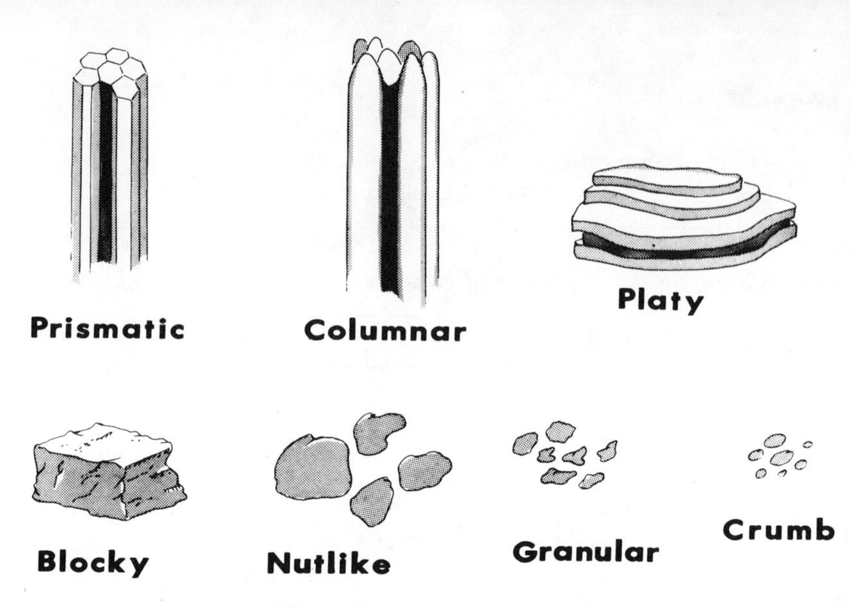

Figure 2-8: Structural Units of Soil. While soil texture involves the size of individual soil particles, soil structure involves the various ways in which these particles may be arranged in groups (aggregates). You can distinguish different soil structures by observing the cracks or "lines of cleavage" of a soil.

growth of a specified plant or crop. A soil in good tilth is well-granulated, which makes it easily cultivated.

Three properties of soil structural units are (1) their shape and size, (2) their strength, and (3) their distinctness.

Shape and Size of Soil Structural Units. When a clod of soil is broken, evidence of soil structure can usually be observed. Careful observation will reveal fine cracks or lines of cleavage. When the cracks are mostly vertical the soil has a prismatic or columnar structure; when cleavage is mostly horizontal, the soil has a platy structure; and when the cracks or cleavage are about equal in all directions, the soil has blocky, nutlike, granular, or crumb structure.

Size of soil aggregates varies with their shape. Blocky or granular structural units usually vary between 0.1 and 5 millimeters in diameter, while the thickness of platy units is of comparable size. Prismatic structural units are usually between 1 and 10 centimeters across and have a greater length than width.

Strength. Strength of soil aggregates is measured by their resistance to crushing.

Distinctness. Distinctness is indicated by the ease with which the structural units can be separated from each other.

Soils are said to be *structureless* when the particles of coarse soil fail to cling together, when fine soil breaks into large clods, or when the soil is *massive*, a single compacted substance.

Soils with good structure provide for a favorable movement of soil air and water; they have more easily available plant nutrients, and are easier to till.

Great Soil Groups of the United States

The soils in the United States are divided, according to their mineral content and other characteristics, into two large soil groups: (1) the *pedalfers*, and (2) the *pedocals*. The pedalfers are found in the eastern half and the pedocals in the western half of the United States.

The word *pedalfer* is derived as follows: *ped* comes from Greek and Latin words meaning "foot" or "soil," *al* comes from *aluminum*, and *fe* comes from the Latin word *ferrum* meaning "iron." Such soils, then, are high in aluminum and iron. They are formed in regions of high rainfall.

The word *pedocal* is derived as above except that *cal* refers to the Latin word *calx*, meaning "lime." Pedocal soils are high in limestone (calcium carbonate) and are formed in regions of low rainfall.

In pedocal soils, most of the water that enters the surface again leaves the surface through evaporation or transpiration. In pedalfers the water moves down into the soil and out through the natural underground drainage system.

The United States Department of Agriculture classification system for soils within these great groups is based on the soil-forming processes. In this system, soils are classified beginning at the lowest unit, the *type*, and progressing to the *order*.

Common usage of such terms as *soil type*, *soil series*, and *soil class* in the classification of soil makes an understanding of the meaning of these terms desirable. Soils which have all horizons similar in distinguishing characteristics, arrangement, and texture, and which have been formed from the same parent material, are placed in the same group and called a *soil type*.

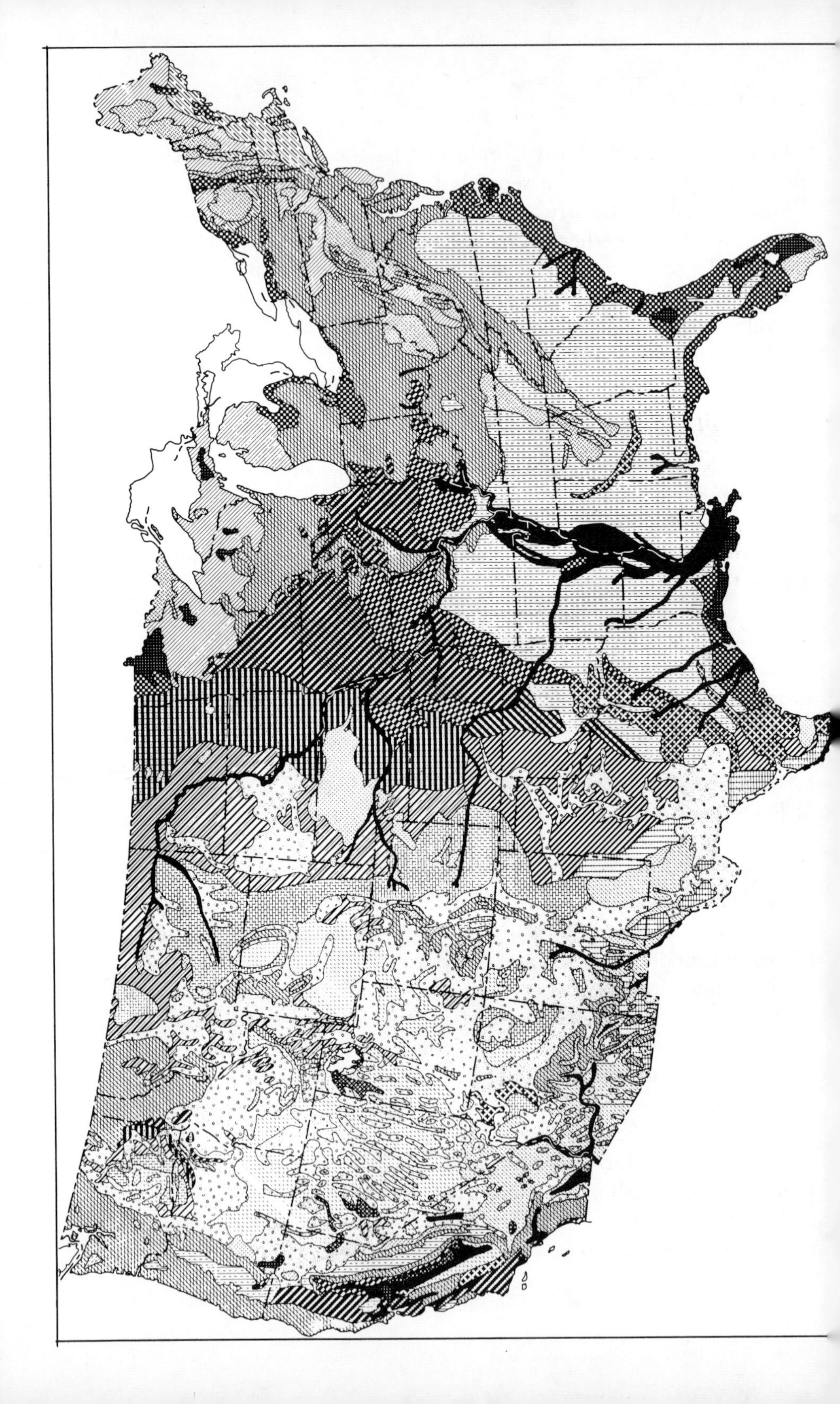

GENERAL PATTERN OF GREAT SOIL GROUPS

The areas of each great soil group shown on the map include areas of other groups too small to be shown separately. Especially are there small areas of the azonal and intrazonal groups included in the areas of zonal groups.

ZONAL

Great groups of soils with well-developed soil characteristics, reflecting the dominating influence of climate and vegetation. (As shown on the map, many small areas of intrazonal and azonal soils are included.)

PODZOL SOILS
Light-colored leached soils of cool, humid forested regions.

BROWN PODZOLIC SOILS
Brown leached soils of cool-temperate, humid forested regions.

GRAY-BROWN PODZOLIC SOILS
Grayish-brown leached soils of temperate, humid forested regions.

RED AND YELLOW PODZOLIC SOILS
Red or yellow leached soils of warm-temperate, humid forested regions.

PRAIRIE SOILS
Very dark brown soils of cool and temperate, relatively humid grasslands.

REDDISH PRAIRIE SOILS
Dark reddish-brown soils of warm-temperate, relatively humid grasslands.

CHERNOZEM SOILS
Dark-brown to nearly black soils of cool and temperate, subhumid grasslands.

CHESTNUT SOILS
Dark-brown soils of cool and temperate, subhumid to semiarid grasslands.

REDDISH CHESTNUT SOILS
Dark reddish-brown soils of warm-temperate, semiarid regions under mixed shrub and grass vegetation.

BROWN SOILS
Brown soils of cool and temperate, semiarid grass-lands.

REDDISH BROWN SOILS
Reddish-brown soils of warm-temperate to hot, semiarid to arid regions, under mixed shrub and grass vegetation.

NONCALCIC BROWN SOILS
Brown or light reddish-brown soils of warm-temperate, wet-dry, semiarid regions, under mixed forest, shrub, and grass vegetation.

SIEROZEM OR GRAY DESERT SOILS
Gray soils of cool to temperate, arid regions, under shrub and grass vegetation.

RED DESERT SOILS
Light reddish-brown soils of warm-temperate to hot, arid regions, under shrub vegetation.

INTRAZONAL

Great groups of soils with more or less well-developed soil characteristics reflecting the dominating influence of some local factor of relief, parent material, or age over the normal effect of climate and vegetation. (Many areas of these soils are included with zonal groups on the map.)

PLANOSOLS
Soils with strongly leached surface horizons over claypans on nearly flat land in cool to warm, humid to subhumid regions, under grass or forest vegetation.

RENDZINA SOILS
Dark grayish-brown to black soils developed from soft limy materials in cool to warm, humid to subhumid regions, mostly under grass vegetation.

SOLONCHAK (1) AND SOLONETZ (2) SOILS
(1) Light-colored soils with high concentration of soluble salts, in subhumid to arid regions, under salt-loving plants.
(2) Dark-colored soils with hard prismatic subsoils, usually strongly alkaline, in subhumid or semiarid regions under grass or shrub vegetation.

WIESENBÖDEN (1), GROUND WATER PODZOL (2), AND HALF-BOG SOILS (3)
(1) Dark-brown to black soils developed with poor drainage under grasses in humid and subhumid regions.
(2) Gray sandy soils with brown cemented sandy subsoils developed under forests from nearly level imperfectly drained sand in humid regions.
(3) Poorly drained, shallow, dark peaty or mucky soils underlain by gray mineral soil, in humid regions, under swamp-forests.

BOG SOILS
Poorly drained dark peat or muck soils underlain by peat, mostly in humid regions, under swamp or marsh types of vegetation.

AZONAL

Soils without well-developed soil characteristics. (Many areas of these soils are included with other groups on the map.)

LITHOSOLS AND SHALLOW SOILS
(ARID-SUBHUMID)
(HUMID)
Shallow soils consisting largely of an imperfectly weathered mass of rock fragments, largely but not exclusively on steep slopes.

SANDS (DRY)
Very sandy soils.

ALLUVIAL SOILS
Soils developing from recently deposited alluvium that have had little or no modification by processes of soil formation.

Figure 2-9: Great Soil Groups of the United States. What is the common soil of your community? Of your state?

A *soil series* is a group of these types which are similar in every respect mentioned above except for the texture of the surface soil, which may vary. The name given to a soil series is generally taken from a nearby town, city, or county where the series is first mapped; for example, the "Webster series" was first mapped in Webster County, Iowa.

The name of a *soil type* is obtained by adding the *soil class* name, such as "clay loam" or "silty clay loam," to the name of the series, as "Webster silty clay loam." (Series plus class equals soil type.) A soil type may vary sufficiently in the amount of stone and gravel, degree of erosion, and slope to warrant further distinction. This variation is known as *soil phase;* for example, "Cecil clay loam, eroded phase."

For the purpose of studying soils and for associating their characteristics and use, these smaller groupings are combined to form many broad soil groups. Soils of these broad groups are distinguished principally by their color and the natural vegetation under which they are formed.

Pedalfers

Iron and aluminum soils, found in the Eastern states, are classified as follows.

Podzol Soils. Podzol soils are developed in cool, temperate, humid climates under the influence of evergreen, deciduous (leaf), or mixed forest vegetation. These soils are characterized, in undisturbed areas, by a surface mat of partially decayed leaves and pieces of wood over a light gray leached soil layer averaging a few inches thick, with or without a thin, dark gray mineral-humus horizon between. The upper subsoil or "B^1" horizon is brown or dark brown and somewhat heavier-textured than the surface soil, and grades through the yellowish-brown, moderately heavy, lower "B" horizon to the parent material. The solum is usually less than three feet thick.

Most of the podzols are strongly acid and have a low natural productivity for cultivated crops, but those having a texture as heavy as sandy loam or heavier may be limed, fertilized, and used for general farming or for grass and other crops in support of dairying. Many podzol soils are either too stony or too sandy for profitable farming and are best used for forests.

Figure 2-10: Pedalfer Soil. Pedalfers are developed in regions of high rainfall under dense vegetation such as the forest in the background. Rain water moves down through these soils into the natural underground drainage system.

Gray-Brown Podzolic Soils. Gray-brown podzolic soils, found in the eastern and midwestern parts of the United States, are developed under deciduous forests in a humid, temperate climate. They lie between the podzols of the North and the red and yellow soils of the South, and join with the prairie soils of the West. Undisturbed soils of this group have a thin layer of leaves lying above an inch or two of dark, grayish-brown, granular humus which grades into a grayish-brown leached horizon, eight to ten inches in depth. The "B" horizon varies from yellowish-brown to light reddish-brown, and is heavier in texture than the surface soil. This horizon fades in color with depth, and the texture grades gradually into that of the parent material. Soil reaction is usually medium acid.

In the Midwestern states, gray-brown podzolic soils were developed largely on glacial till. The productivity of these soils varies considerably according to the texture and the nature of the parent material. In general, the soils developed on glacial till are more productive than those developed on sand. Large areas have been cleared and are used for general farming, and sandy soils near cities are used for truck farming. These soils are responsive to liming and to the application of organic matter and mineral fertilizers.

Red and Yellow Podzolic Soils. The red and yellow podzolic soils occupy an extensive area in the southeastern part of the United States. The topography varies from flat to rough and hilly. The climate is warm-temperate to subtropical, and humid. The native vegetation is largely forest—pines, and oaks and other hardwoods. The soils are strongly leached, acid in reaction, and low in organic matter and mineral fertility. The surface soils are light-colored and sandy, and the subsoils are heavier, tougher, and of red, yellow, or mottled color. Though low in fertility, these soils are easily tilled and respond readily to fertilization.

Prairie Soils—Northern and Southern. The true prairie soils have developed in cool, moderately humid climates, under the influence of grass vegetation. These soils occur in the Central states and occupy a large part of the Corn Belt. The profile is characterized by dark-brown to nearly black, slightly acid surface soil underlain by brown, well-oxidized subsoil. The parent materials have a wide range of composition, especially in their content of lime. This group includes soils that rank among the best in the United States for farming.

The *reddish prairie soils* lie south of the region of the true prairie soils. They have a redder color because under higher temperatures the iron tends to be better oxidized and the organic matter content is lower. They are for the most part productive soils, and much of this land is cultivated.

Pedocals

Calcium soils are found in the Western states and are categorized as follows.

Chernozem Soils. Soils of the chernozem group are developed in temperate, subhumid grasslands. The surface soil is very dark brown to black. The parent material varies greatly, but a distinguishing feature is an accumulation of calcium carbonate in

Figure 2-11: Pedocal Soil. This Montana rancher and his daughter live in a region where limited rainfall keeps the soil high in lime content.

the lower part of the solum, which characterizes all pedocal soils. These soils were developed from the eastern part of the Dakotas to central Kansas on the tall-grass plains where the rainfall ranges from 18 to 30 inches. Wheat and other small grains are grown in this area.

Dark Brown (Chesnut) Soils—Northern and Southern. Northern dark brown or *chestnut soils* occupy a large area in the Great Plains of northeastern Montana, the western Dakotas, and Nebraska. They are dark brown or dark grayish-brown, and grade into light gray or white calcareous horizons at a depth of 1½ to two feet. They develop in temperate to cool semiarid regions under mixed short and tall grass. The main crops are small grains, mostly spring and winter wheat. With adequate moisture, they are highly productive, but average yields are low as a result of insufficient rainfall.

The similar, Southern dark brown or *reddish chestnut soils* are found on the grassy plains from southern Kansas through Oklahoma and Texas to the Gulf of Mexico.

Brown Soils. Brown soils cover a vast area in the western part of the Great Plains and smaller areas in the intermountain country of the Far West. These soils exist under a temperate or cool semiarid climate and a native vegetation of short grasses, bunch grasses, and shrubs. The surface soils are brown, and the subsoils grade at depths ranging from one to two feet into light gray or white calcareous layers. Much of this land is used as livestock range, though large areas are dry-farmed, mostly in small grains. Irrigated areas produce a wide variety of crops.

Reddish-Brown Soils. Reddish-brown soils are in semiarid areas of the Southwest from western Texas to southern Arizona. These soils exist under a warm-temperate climate with hot summers, and support thin growths of short grass or bunch grass with scattered shrubs and small trees. The surface soils are reddish-brown to red in color and of mellow consistency; the upper subsoils are red or reddish-brown, heavy, and tough; and the lower subsoils are pink or nearly white and very limy. Most of the land is in livestock ranches or public range, though some areas are irrigated and highly productive.

Gray Desert Soils. Desert soils occupy vast reaches of desert or semi-desert in the Western intermountain region. The climate is arid and warm to cool-temperate. The vegetation is mostly desert shrubs. The surface soils are typically light grayish-brown or gray in color, and low in organic matter. Subsoils are slightly lighter in color and high in lime. These soils are leached very little and thus are rich in mineral nutrients, but in places they contain very high concentrations of soluble salts. Most of the land is useful only for livestock range with low carrying-capacity, but under irrigation it is highly productive.

Summary

Soil is the mineral and organic material that supports plant growth on the earth's surface. It is a mixture of particles of rock, organic materials, living forms, air, and water.

Soil is developed by the process known as *weathering* from igneous, sedimentary, and metamorphic rock; partially decomposed rock is known as *parent material.* There are two classes of organic parent material also, peat and muck. Most parent material is moved from the place of its origin by (1) water, (2) wind, (3) gravity, or (4) ice and glaciers.

Soil lies in layers called *horizons,* classified according to the amount of weathering and decay that has taken place. The top, most "finished" layer, roughly corresponding to the topsoil, is called the *"A" horizon.* Below it lies the *"B" horizon,* roughly corresponding to the subsoil. Together the "A" and "B" horizons form the "true soil" or *solum.* Below these lies the *"C" horizon,* which consists of parent material.

Soils are differentiated by (1) color, (2) texture, and (3) structure.

The dark color of topsoil usually is produced by organic matter. Subsoil color usually is determined by the presence of iron in various characteristically colored forms.

Soil texture is the property of soil controlled by the size of individual grains or particles. The three major size particles, from largest to smallest, are sand, silt, and clay. Loam is a soil having relatively equal proportions of the three types.

Soil structure refers to the grouping or arrangement of soil particles. A single mass or cluster of particles is called a *soil aggregate. Soil tilth* refers to the fitness of a soil for cultivation. The properties of soil structural units are their shape and size, their strength, and their distinctness.

The soils of the United States are divided into two great groups, the pedalfers and pedocals. Pedalfers are Eastern soils, and are high in aluminum and iron. They are found in regions of high rainfall. Pedocals, found in regions of low rainfall, are Western soils, and are high in limestone (calcium carbonate).

Soils which have all horizons similar in characteristics, arrangement, and texture and which have been formed from the same materials are a *soil type.* The *soil series* is a group of soil types which are alike in every respect above except that the texture or depth of the topsoil may vary. The name of a soil series is taken from a town or city near where it was first mapped. The name of a soil type is obtained by adding the soil class name ("clay loam") to the name of the series.

Study Questions

1. What is soil?
2. How is soil formed?

3. What is a soil profile?
4. What determines soil texture?
5. What is soil structure?
6. What are the materials and processes of soil development?
7. How are soils classified?

Class Activities

1. Refer to the map of great soil groups of the United States to identify the major soil group of your community, and study the description in the text.
2. Collect samples of soil from home farms and classify them as to soil texture. First determine the amount of (1) sand, (2) silt, and (3) clay in the soil by allowing it to settle in a jar of water after being shaken vigorously. Use the textural chart to determine the textural class.
3. Identify the kinds of parent materials that are to be found in local soils.

Organic Matter in Soil

All dead plant and animal materials that remain in the soil make up its organic matter. Since the beginning of time, farmers have used one form of organic matter, barnyard manure, to increase the productivity of their soils. This chapter is intended to develop a better understanding of the major role of such materials in increasing soil productivity.

Importance of Organic Matter

Soil, as we have seen, is a mixture of weathered rock materials, organic matter, living forms, air, and water. Soils rich in organic matter have better tilth and productivity than those containing chiefly weathered rock materials.

Soil erosion, along with continuous cropping, has lowered the organic content of many soils. They are kept productive only by use of large amounts of commercial fertilizers, improved seed, insecticides, and weed control chemicals.

What Is Organic Matter? All plant and animal residues in the soil constitute organic matter. Besides crop residues, barnyard manure, green manure, and dead roots, this includes the decomposed carcasses and excrements of insects, worms, and larger animals. Another important part includes dead soil microorganisms such as bacteria, fungi, and protozoa.

Chemically, organic matter contains (1) carbohydrates, (2) proteins, and (3) fats, resins, waxes, and similar compounds.

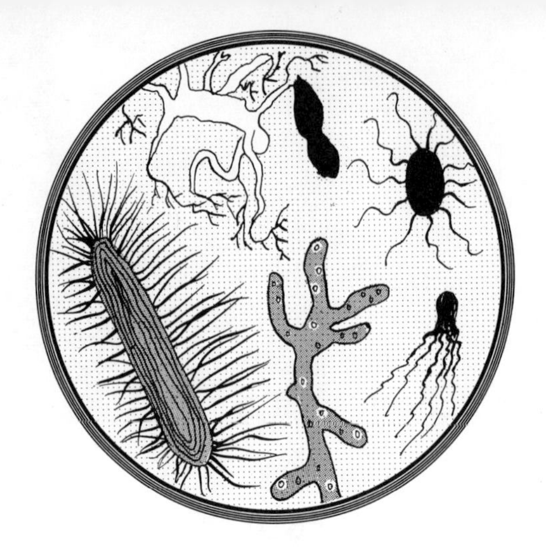

Figure 3-1: Microorganisms. Microscopic plants and animals in the soil break down organic matter and make the nutrients it contains available to plants.

Mineralization is the process by which organic matter is broken down into simple mineral compounds. The products of this breakdown are water, carbon dioxide, free nitrogen, ammonia, methane gas, and a few simple mineral salts.

All dead organic material added to the soil finally breaks down through the process of mineralization. Some organic residues break down rapidly, while others, such as lignin, are more resistant. Lignin, which in addition to cellulose constitutes the woody parts of plant tissues, is (with proteins) the major constituent of humus.

Humus. Soil scientists speak of *humus* as well as of *organic matter* because humus is a relatively stable stage in the breakdown of organic matter. It is the product of many sources of organic matter and is usually recognized by its brown to black color. Humus is practically odorless. While it has no definite chemical composition, it is a uniform type of material. It has no particular form, and it represents a relatively small percentage of the total soil volume.

Manures and other forms of organic matter, in breaking down into humus, release their nutrients for crop use. Soil microorganisms, such as bacteria and fungi, are the active agents responsible for this breaking down, the raw organic materials being a source of food for the living soil organisms.

Humus has some of the same physical properties as clay particles. It can absorb considerable moisture and thus adds to the moisture-holding capacity of soil. Soil chemical elements become attached to humus particles. Humus can bring about *crumb formations* of soil particles and thus improve soil tilth. However, it is not sticky like clay when wet and helps to lessen the stickiness of clay soils.

The humus content of soil has been built up over a long period of years and is fairly constant unless top soil is carried away by forces of soil erosion. Plant residues decay to humus within a comparatively short period of time—from several months to a year or two. An exception to this is the slow rate of decomposition of plant materials in peat soils. Peat soils are wet, poorly aerated, and usually acid. All such conditions are unfavorable for soil microorganisms, which are the principal forces bringing about decay of organic matter.

Humus and Nitrogen Content of Soil. The amount of nitrogen in the soil is almost in direct proportion to the amount of humus in the soil, because nitrogen is largely stored in the humus. In other words, the percentage of nitrogen in the soil is an index to the percentage of humus and organic matter. Land appraisers often rate the productivity of the soil according to the percentage of nitrogen in it.

Organic Matter and Soil Tilth. Farmers speak of soils that are easily worked as having good soil *tilth.* Soils that pack are thought to have lost their tilth. Soils rich in organic matter generally have good tilth.

Soils scientists describe soils with good tilth as *friable,* which means that they are easily crumbled or pulverized. The soil particles are held together in small crumbs or granules.

Figure 3-2: Eroded Pasture. Humus is carried away with topsoil when erosion occurs.

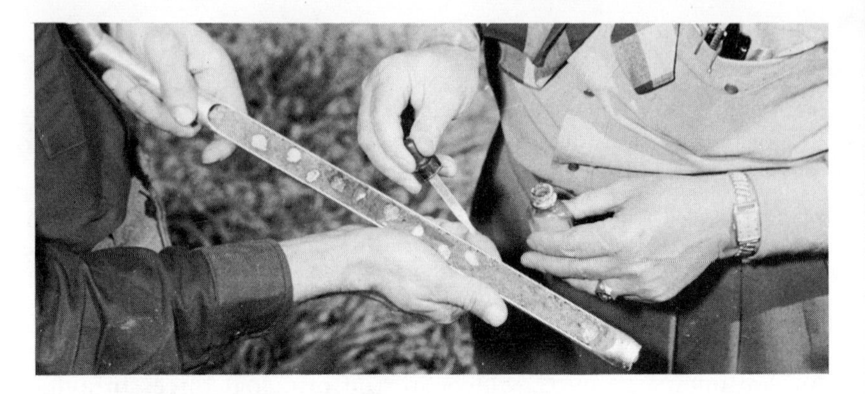

Figure 3-3: Testing Tilth. Loose soils, which have good tilth, absorb a white chalk solution. The solution remains on top of tight soils.

Farmers generally test tilth, before plowing or cultivating, by squeezing a handful of moist soil. If the soil remains in a hard lump it has poor tilth, and hard clods will result from plowing. Soils with good tilth are loose, mellow, and well-granulated.

Soils of poor tilth are sometimes called "noontime soils." They are too wet to work before noon and too dry to work after noon. Soils of good tilth can be worked earlier after a rain.

Organic matter promotes granulation by binding the soil particles into granules. Adding organic matter to heavy soils makes them less sticky. Soils with good tilth can absorb and hold moisture. Deeper root growth takes place in soils with good tilth. Rain and irrigation water have difficulty penetrating soils with poor tilth.

Soils with good tilth permit air to enter the soil. Air is needed by soil microorganisms and plant roots. Sandy soils having too much aeration, however, tend to dry out. Organic matter added to sandy soils will lessen excess aeration and increase moisture-holding capacity. Organic matter helps bind the particles of sandy soil together.

Soils with good tilth are ready to be tilled earlier in the spring than soils with poor tilth. Sandy soils are generally first to warm up.

Continuous cultivation and planting of row crops have lowered the tilth of many soils, because much of the organic matter and humus in the soil has been "burned out" or used up. Some soils which at one time had good drainage now need to be tiled as a result of a loss of soil tilth. Subsoils exposed by erosion often have poor tilth because they lack organic matter. Field spots showing lighter subsoil thus usually have poor tilth. Sandy soils become more subject to drought from loss of organic matter and the consequently lower water-holding capacity.

Cultivated soils with good tilth are less subject to water and wind erosion. The surface soil is held together in granules by organic matter. Water penetrates these soils better and the granules are more resistant to the forces of wind and water than smaller soil particles.

Maintaining and Increasing Organic Matter

The difficulty of increasing the percentage of organic matter in the soil is great. A better policy is to maintain the percentage already contained in the soil. Mineral soils contain an average of 2 to 4 per cent of organic matter, about 20 tons per acre. It would take 40 years to double this percentage. Soil management practices thus should be planned to maintain organic matter as well as to use the soil most productively.

The roots of plants are the largest source of natural organic matter. Other sources are barnyard manure, crop residues, and green manures. Only a part of added organic matter stays in the

Figure. 3-4: Roots Beneath Sod. Deep-growing roots like these are the largest source of natural organic matter.

Figure 3-5: Rotation for Soil Maintenance. Grass crops grown in a rotation help maintain organic matter in the soil.

soil as humus. A larger part of it disappears during decay as carbon dioxide gas and other decomposition products. Under some conditions, almost all of a green manure crop may be lost in the form of carbon dioxide gas. Under average conditions, however, no more than half of the green manure may be lost, and the remainder will be retained as part of the soil humus for a limited time.

Grassland Farming Increases Soil Organic Matter. Grassland farming conserves and increases the organic matter content of a soil. Decaying grass roots keep up the organic content of soils, while cultivation speeds the loss of organic matter. This loss takes place because of greater oxidation of organic matter and the increased activity of soil organisms. Thus, cultivated fields produce crops at the expense of organic matter.

Consider what happens when virgin peat soils are drained and cultivated. Uncultivated peat soils are high in organic matter, which has been accumulated over a period of years. Peat soils are normally wet, poorly aerated, and acid, all of which conditions slow the process of decay. When peat fields are drained and cultivated, decomposition frequently becomes active enough to actually lower the land surface.

Cultivation also speeds decomposition of organic matter in mineral soils.

Grass crops in a rotation build up organic matter. Thus the amount of organic matter in the soil is more nearly maintained at its original level.

Legumes and Non-Legumes As Green Manure Crops. Legumes add both nitrogen and organic matter to soils, while non-legumes only add organic matter. Bacteria, which are largely responsible for the decay of organic matter, must have access to nitrogen. In the decomposition of mature non-legume manures, bacteria draw upon the soil nitrogen they need while decomposing the organic material. As a result, while green manure crops such as rye can add a large quantity of organic matter to the soil, the crop following may be starved for nitrogen. A good practice would be to add nitrogen fertilizer to soils fertilized with non-legume green manure crops.

The Carbon-Nitrogen Ratio. Soil scientists stress the importance of maintaining a desirable *carbon-nitrogen ratio.* A major element in plant residues is carbon, found in starches, carbohydrates, proteins, and fats. Nitrogen is contained in plant proteins. The normal carbon-nitrogen ratio is ten pounds of carbon to one pound of nitrogen. Legumes and young grass residues added to the soil have a nitrogen-carbon ratio of 20 or 30 to 1. However, mature hay, such as timothy, could have a carbon-nitrogen ratio of 50 to 1 and wheat straw a ratio of 60

Figure 3-6: Green Manure. When a green manure crop is plowed under, organic matter in the soil is increased.

Figure 3-7: Crop Residues. To prevent an unfavorable carbon-nitrogen ratio, many farmers add nitrogen fertilizer to soil after plowing in large amounts of non-legume crop residues.

or 80 to 1. Adding hay straw and corn stalks to the soil will temporarily change the carbon-nitrogen ratio.

The importance of the carbon-nitrogen ratio is better appreciated when one understands what takes place in the soil when crop residues are added. The decay of crop residues is largely brought about by soil microorganisms, such as bacteria, fungi, and actinomycetes. These organisms multiply in huge numbers as new organic matter is added to the soil for them to feed upon, but they require a balanced diet of nutrients, much as do farm crops.

Crop residues such as grain straw do not contain enough nitrogen to feed soil microorganisms. In breaking down the crop residues in such a situation, soil microorganisms must draw upon the nitrogen supply in the soil. The available nitrogen in the soil becomes temporarily tied up, and a short supply of nitrogen for crop use results. Soils in which this situation occurs are said to have a *temporary nitrogen depletion.*

An approved practice is to add nitrogen fertilizer to soils to which considerable amounts of straw residues have been added. The nitrogen fertilizer will aid the soil microorganisms in speeding up the decay of the new organic matter. Straw residues kept as a surface mulch, rather than plowed under, will lessen the unbalance between carbon and nitrogen. However, the straw decomposition will be slower.

Compost. Many gardeners maintain a compost pile as a source of organic matter. Compost piles are a mixture of manure, soil, lime, fertilizer, straw, old hay or grass, peat, muck, and similar materials. The addition of the manure, soil, lime, and fertilizer brings about a quicker rotting of the straw and other organic materials. Barnyard manure is generally used in alternate layers with straw, leaves, sod, and similar materials. The material to be composted is spread out and piled in layers six inches to one foot in thickness. Each layer is sprinkled with fertilizer and lime and thoroughly wetted. Compost piles are usually built up to a height of four to six feet. The mass is kept moist and is occasionally turned. Compost, sometimes called "artificial manure," can be prepared in three to four months by this means.

Summary

All dead plant and animal materials that remain in the soil make up its organic matter. Soils rich in organic matter have better tilth and productivity.

Soil is of good tilth when it is easily worked and crumbles into soil granules which permit soil aeration and water penetration. Soils on many farms have lost much of their tilth as a result of a loss of organic matter. Farmers who can improve tilth will improve their soil workability and drainage, and lessen the danger of soil erosion.

Humus is a relatively stable stage in the breakdown of organic matter. The amount of nitrogen in a soil is almost in direct proportion to the amount of humus. Land appraisers often rate soil productivity by the amount of nitrogen it contains.

It is difficult to increase the percentage of organic matter in the soil. Good soil management practices thus should be planned to maintain and increase the present proportion of organic matter.

The roots of plants are the largest source of natural organic matter. Other sources are barnyard manure, crop residues, and green manures. Grassland farming conserves and increases the organic matter in soil, while cultivation depletes it. Legumes as green manure crops add both nitrogen and organic matter. Non-legumes as green manure crops only add organic matter. It is a good practice to add nitrogen to soils fertilized with non-legume green manure crops or with mulches. This maintains a desirable carbon-nitrogen ratio. Many gardeners maintain a compost pile as a source of organic matter.

Study Questions

1. Why is organic matter so important?
2. How is humus formed and why is it so important?
3. Why is the ratio of carbon to nitrogen important to crop production?
4. How is organic matter lost by continuous cultivation and cropping?
5. Is the nitrogen content of soils an index to soil fertility?
6. Why are legumes better green manure crops than non-legumes?
7. How does grassland farming maintain soil organic matter content?
8. What is good soil tilth and what is its value?
9. How would you make a compost pile?

Class Activities

1. Collect samples of soil from local farms and compare their tilth.
2. Conduct experiments on the water-holding capacity of soils containing varying amounts of organic matter.
3. Conduct class demonstrations in the effect of water runoff on soils with varying amounts of organic matter.
4. Discuss the practices of local farmers that aid or lower the organic matter content of soils.
5. Observe soils that have been under grassland farming and compare them to soils that have been used for continuous row cropping.
6. Prepare a compost pile.

Plant and Animal Life in the Soil

Many farmers are not aware of the important role played by plant and animal life in the soil. A handful of soil may contain as many living organisms as there are people in the world. The largest number of these organisms are bacteria. They can be seen only under a microscope. In addition to roots, the observable kinds of life in the soil include earthworms and beetles and their various kinds of larvae. Both mice and moles are common in undisturbed soils of forests, pastures, and range lands, and mice are common in hay and grain fields. Larger inhabitants of soils are gophers, woodchucks, and prairie dogs.

True soil with the important life-giving humus of decayed organic matter is largely the product of soil microorganisms. Soil tilth is associated with the good physical condition of the soil, which is influenced in a large degree by these minute forms of plant and animal life. Many plant nutrient elements would be unavailable for plant use if it were not for the action of soil bacteria and other microbes.

Soil bacteria perform the spectacular feat of taking free nitrogen from the soil air and combining it with other elements into a usable form for higher plants. Farm crops are unable to draw upon the vast stores of nitrogen in the air until bacteria and other forces have changed it into a form that can dissolve in water.

Few farmers have fully realized that they need to grow plant life beneath the surface of the soil as well as on top of it. Such growth involves both the roots of farm crops and the billions of minute soil organisms which make the soil productive.

Figure 4-1: Crop Growth and Microorganisms. This photograph dramatically illustrates the value of microorganisms in soil. The fine growth of peas at left was aided by treatment of the seed with nitrogen-fixing bacteria; the peas at right were untreated.

Microorganisms

To know our soils, we first must know what kinds of microscopic plant and animal life are in them. Much of our knowledge of such life in the soil must be credited to our bacteriologists and zoologists. These scientists have developed means of identifying soil microorganisms.

Bacteria. The most numerous and probably the most important soil organisms are bacteria. Bacteria are single-celled plants which under favorable conditions can multiply in large numbers. They can reproduce in as short a time as 20 or 30 minutes. Under unfavorable conditions, however, the number of bacteria decreases rapidly.

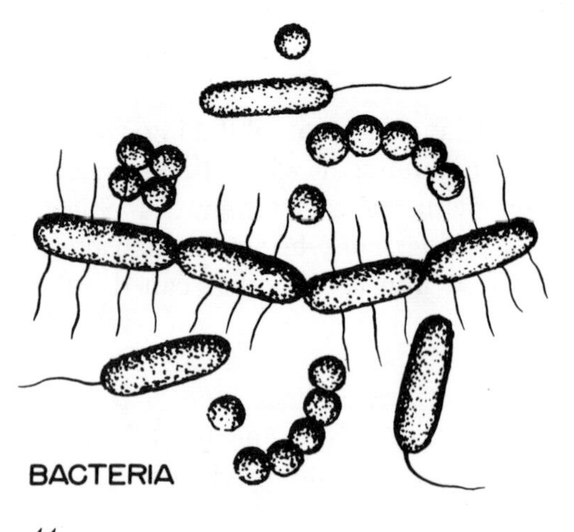

Figure 4-2: Bacteria. Here greatly magnified, bacteria are tiny plants that break down organic matter and minerals into forms that can be used by crops.

A soil containing 5 per cent organic matter may contain up to 3½ tons live weight or 0.7 tons dry weight of bacteria per acre.

Farm crops require free oxygen or aerated conditions in the soil. Some bacteria called *anaerobic bacteria* can live under conditions in which free oxygen is absent. Anaerobic bacteria can get their oxygen through chemical reactions of such items as sugar, which contains oxygen molecules.

Actinomycetes. Actinomycetes are about as large as bacteria, but some resemble molds in their manner of growth. Actinomycetes can live under drier conditions than bacteria, and are generally abundant in sod. They are among the most important agents of the soil in the breakdown of dead plant materials, including cellulose, which is more resistant to bacterial action. The characteristic smell of newly plowed land in the spring is thought to be due to odors cast off by actinomycetes.

Fungi. Many of the soil fungi are too small to be seen by the naked eye, though mushrooms, puffballs, and toadstools are the fruiting part of one group of soil fungi. Molds, a form of fungi, are observable forms of woolly growth. Yeasts and yeast-like fungi are common on the skins of fruit and in orchard soils. Fungi do not have chlorophyll and must secure their food from organic substances.

Algae. Soil algae differ from bacteria in that they contain chlorophyll, the green coloring matter in plants. Chlorophyll bodies (chloroplasts), under the stimulus of light, are active in

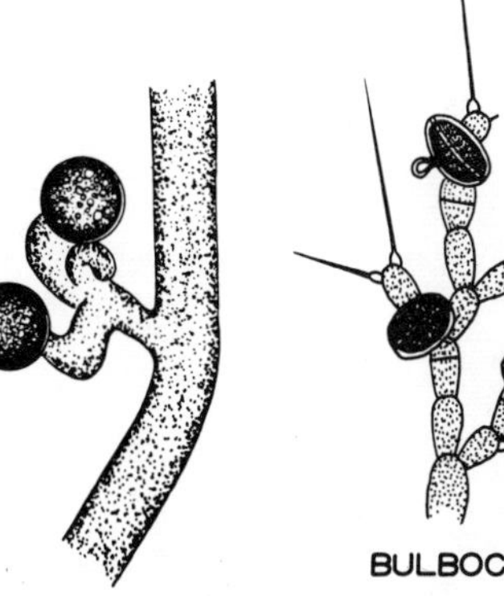

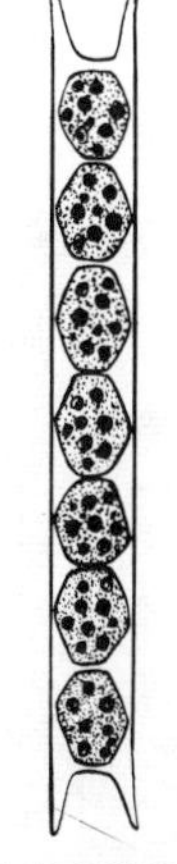

Figure 4-3: Fungi and Algae. Fungi and algae, here magnified, also enrich the soil.

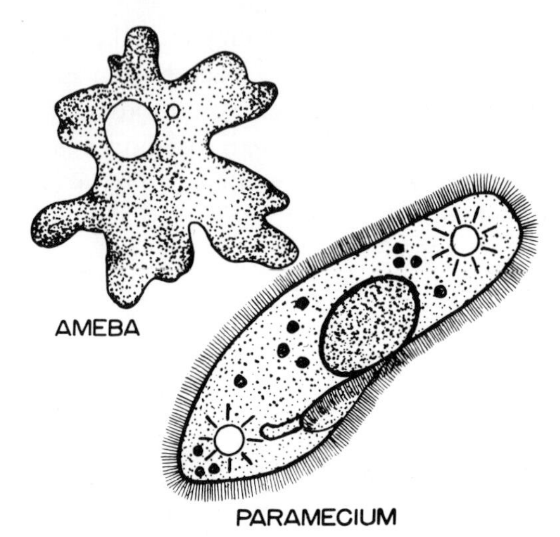

Figure 4-4: Protozoa. This magnified view shows two kinds of tiny animals that live in wet soil. Note the tiny hairs (cilia) that propel the paramecium.

the manufacture of sugar and starch from carbon dioxide and water. Among the larger, visible forms of algae are seaweeds and scum-forming algae of ponds. Soil algae are too small to be observed by the naked eye, though large numbers may give a green color to the surface of a soil. They are generally found on or in the first few inches of soil. Some soil algae are single-celled organisms. Others are in the form of simple filaments or colonies.

Algae favor damp soils exposed to the sun and are most active in the spring and autumn. They may add some organic matter to the soil, and help bind soil particles together. They are believed to be able to fix free nitrogen from the air in the soil. Algae are

Figure 4-5: Fungi in Sod. The dense green arc in this lawn is part of a "fairy ring." Heavy surface growth has been caused by the activity of fungi within the soil.

also thought to be of value in rice production under water-logged conditions by supplying oxygen in addition to nitrogen.

Protozoa. Protozoa are minute animals found in the soil. They are the smallest form of animal life and are single-celled. They reproduce by dividing. They are usually mobile, having whiplike flagella or body hairs (cilia) which can propel them. Protozoa are found in large numbers in stagnant waters. Even in soil, they must live in a water film. When the soil dries the protozoa become inactive. Protozoa feed on soil bacteria and are found in large numbers in many soils.

How Soil Microorganisms Help Farmers

The activities of soil microorganisms are beneficial in many ways. Decay of plant residues such as straw would not be possible without them. The vast amount of free nitrogen in the air is made available to plants by the work of nitrifying bacteria. Wonder drugs such as penicillin have been produced by soil microorganisms. An understanding of these and the other benefits discussed below is essential to good soil management.

Plant Decay. The life processes of soil microorganisms are the most vital force in the rapid decay of dead plant materials. Feeding on these dead plant materials, the microorganisms break down the sugars, starches, cellulose, lignin, proteins, and fatty substances of plant tissues, and excrete as waste products the carbon dioxide, ammonia, and minerals needed by plants for growth. The breakdown of lignin by bacteria is especially important, because lignin is a major constituent of organic matter. That portion of the dead plant materials which has been absorbed as food by the microorganisms is also added to the soil organic matter when the microorganisms die.

Releasing Mineral Elements. Much of the plant food in the soil is not available for plant use. It takes the action of soil microorganisms and other forces to release nutrients as they are needed. As straw and other organic materials are decomposed, large quantities of carbonic acid are produced. This acid has a solvent action on soil minerals. Quantities of nitric and sulfuric acid are similarily produced. These acids also help change minerals from an insoluble to a soluble form. For instance, calcium phosphate may be changed to superphosphate, and calcium carbonate to

Figure 4-6: Nodules. Each of the lumps on these clover roots contains millions of nitrogen-fixing bacteria.

calcium sulphate. Soil scientists believe that microorganisms may be one of the most important means of making mineral elements available for plant use.

Nitrogen Fixation. Some bacteria can take free nitrogen from the soil air and utilize it in building their cells. Thus, the free nitrogen is brought into a "fixed" form. The nitrogen-rich cell materials of bacteria eventually become a source of nitrogen for the farmer's crops. This process of changing nitrogen from the air into nitrogen compounds by soil bacteria is called *nitrogen fixation.*

Soil scientists believe that some fungi and blue-green algae also can utilize the free nitrogen of the air.

Bacteria which are capable of nitrogen fixation are of two groups. One group is attached to the roots of legumes. The second group live in the soil independent of plants. These are called *free nitrogen-fixing bacteria.*

Some of the free nitrogen-fixing bacteria require neutral soil, while others can live under fairly acid conditions. The legume group requires good aeration, abundant organic matter, ample calcium and phosphorus, plentiful moisture, and favorable temperatures. Lime added to the soil, where needed, benefits the action of these bacteria. Application of phosphate and other mineral elements also aids their activity.

Some soluble nitrogen compounds may be passed out by bacteria during their growth. A definite source of nitrogen for higher plants is the dead bacteria cell materials which are decomposed or broken down into a form which plants can use. This breaking-down process is called *mineralization of nitrogen.*

Nodule Formation. The most important group of nitrogen-fixing bacteria are those which cause nodules or "little lumps" to form on the roots of legumes. These bacteria enter the single-celled root hairs, where they multiply rapidly because of the favorable food situation. They form an infection thread which penetrates the cortex or outer tissue of the root and causes the root to produce numerous cells which eventually form nodules, usually in bunches. A nodule may hold millions of bacteria.

Insects or other forms of bacteria may injure legume roots, causing the formation of false nodules. On cultivated annual legumes *true nodules* are ball-like, while on legumes such as alfalfa and the clovers true nodules are long and clustered.

Plants absorb the nitrogen collected in the nodules. Some may be excreted from the nodules into the soil. The amount of nitrogen in the nodules is a tiny fraction of that fixed.

Inoculation for Legumes. When a legume plant is grown for the first time, it is essential to *inoculate the soil with the proper bacteria.* This is usually done by inoculating the seed at planting time. Inoculation materials can be purchased at most seed stores.

Figure 4-7: Inoculation. When used to inoculate legume seed with nitrogen-fixing bacteria, a commercial preparation is first mixed with water, then applied to the seed as shown.

The required bacteria may live in the soil for 10 to 20 years even without their host plant. A recommended practice, however, is to inoculate seed each time that a new crop is grown. The cost of inoculation is small compared to the possible benefits. Some strains of nitrogen-fixing bacteria may be more effective than others. Commercial inoculation materials usually contain good strains of bacteria.

The following legume groups can utilize the same nitrogen-fixing bacteria:

Group 1: Alfalfa, and white and yellow sweet clover.
Group 2: Red, alsike, crimson, mammoth, and white Dutch clover.
Group 3: Garden and field peas, sweet peas, and vetches.
Group 4: Garden, kidney, and navy beans.
Group 5: Soybeans.
Group 6: Cowpeas, peanuts, lespedeza, velvet beans, and lima beans.

Amounts of Nitrogen Added. Soils most lacking in available nitrogen profit most from the nitrogen-fixing bacteria. Soil scientists estimate the increases in soil nitrogen by the bacteria which live in the soil independent of higher plants to be 10 to 50 pounds per acre per year. All crop plants can profit from this source of nitrogen.

Under favorable conditions, 50 to 100 pounds of nitrogen per acre may be produced in a year by the nitrogen-fixing bacteria attached to legume root nodules. However, there may not be a net gain of nitrogen in the soil as a result of the nitrogen-fixing bacteria—the amount produced may be equal to the amount harvested in the crop. Plowing down a legume green manure crop adds to the total supply of nitrogen in the soil.

Antibiotic Properties of Soil Organisms. Most farmers are familiar with the use of antibiotics in livestock feeding. Soils have antibiotic properties because of substances produced by certain forms of bacteria and fungi. Health-giving antibiotics that have been produced by soil microorganisms are *penicillin* and *streptomycin.* Penicillin is produced by a large number of fungi. Streptomycin is produced by certain actinomycetes.

The antibiotic effect of soil microorganisms helps control diseases which may affect plant growth.

Figure 4-8: Penicillium. This microscopic fungus produces valuable antibiotics, which are chemical substances that kill harmful bacteria.

The discovery that many soil microorganisms can produce antibiotics is among the great medical triumphs of modern times. Scicntists arc scarching for more wonder drugs produced in the same way.

Aiding Soil Tilth. Some soil scientists believe that the by-products of bacteria, fungi, and actinomycetes cement soil grains together, thus improving aggregation. Others believe that soil grains are held together by microbial threads or the cells of soil microorganisms. In any event, conditions which promote the growth of soil microorganisms appear to be the most important factor in the improvement of soil tilth.

Harmful Effects of Soil Microorganisms

The undesirable effects of microbes also need to be recognized. Some cause plant and animal diseases. Others use up or tie up important plant nutrients.

Disease Organisms. The soil is a home for many bacterial diseases, such as *wilt* of potatoes and tomatoes. Some fungi cause *damping-off of seedlings, mildews, blights, wilt,* and *dry rot* of potatoes. Some actinomycetes cause *scab* on potatoes and sugar beets and *pox* on sweet potatoes. Most of these conditions can be corrected by applications of various chemicals.

Competition for Plant Nutrients. Bacteria and other soil microorganisms use about the same food elements as do farm crops. When any of these elements is in short supply, a large population of soil microbes can reduce the amount available for plant use. Soil microorganisms use considerable quantities of nitrogen, phosphorus, oxygen, carbon, and hydrogen. The keenest competition between soil microorganisms and farm crops is for nitrogen. A *temporary loss of nitrogen* may occur when nitrogen is stored in soil microorganisms.

Loss of Free Nitrogen. Heavy, compact soils may lose some nitrogen through the action of soil microorganisms. Such a loss is minor in well-aerated soils. Under conditions in which oxygen is deficient, microbes break down the nitrogenous materials into ammonia or gaseous nitrogen, which escapes from the soil. This process is more active in soils containing soluble organic matter, which serves as a good source of energy for soil microorganisms.

Nitrogen "Tie-Up." When large amounts of fresh plan residues, such as green manure or straw, are added to the soil, the number of soil microbes is greatly increased. This situation will generally lower the available supply of nitrogen for plant growth. Since many crop residues such as straw contain mostly carbon mate-

Figure 4-9: Microorganisms in the Laboratory. Microorganisms isolated from soil samples collected around the world are processed, identified, and stored in this "culture library."

rials, the soil microorganisms draw upon the soil nitrogen to balance their diets. Thus the available soil nitrogen is "tied up" in the tiny bodies of the microbes.

The nitrogen is not released until the bacteria begin to die in large numbers. This takes place rapidly after the crop residues have been broken down.

Legume green manure residues contain a higher percentage of nitrogen than grain straw. Consequently, bacteria acting on legume green manure draw less heavily upon the nitrogen elements in the soil to complete the process of decomposition.

Conditions Favorable for Soil Microorganisms

Factors which regulate the growth and activity of soil microbes are (1) food, (2) temperature, (3) moisture, (4) acidity, (5) aeration, and (6) light.

Food Requirements. The more fertile a soil is, the more microbes it contains. The nutrient requirements of soil microbes are somewhat similar to those of plants that grow above ground. Only certain strains of soil algae contain chlorophyll, however, which aids in the manufacture of sugars and starches. Most soil microbes thus must depend on the organic matter of other plants and animals for sugars and starches. The chief sources of food for soil organisms, then, are crop residues (including roots), other soil organic matter, and soil mineral elements.

Free nitrogen of the air is a source of food for certain bacteria. In the presence of light, soil algae, which contain chlorophyll, can perform the function of manufacturing food. Some bacteria can use the cellulose of organic materials, while others can only use the more digestible sugars and starches.

Soil Temperature. Growth of soil microbes is retarded by both low and high soil temperature, though most of these organisms appear to be able to adjust their activity to slow temperature changes. Good crop-growing conditions generally favor growth and activity of microbes.

The dark color of soils rich in organic matter makes them warm up faster than light-colored soils. Farmers have observed that sandy loam soils warm up earlier in the spring. Undoubtedly, springtime bacterial activity is higher in these loamy soils than in heavy wet soils.

Figure 4-10: Moisture and Microorganisms. Like crops, microorganisms require abundant moisture for healthy growth.

During the winter there is little or no activity by nitrifying bacteria. In the spring, when ground temperatures are yet cool, bacteria start to produce available nitrogen. As soils dry out and the temperatures increase to high levels, the bacterial activity is slowed.

Moisture. Soil microorganisms can adjust themselves to a wide range of moisture conditions. Moisture conditions favorable to crop growth are generally favorable to the growth of soil microorganisms.

Acidity. Beneficial microbes thrive best in soils that are approximately neutral in reaction. Soil reaction is its degree of acidity or alkalinity, expressed in *p*H values. Bacteria generally favor a *p*H range of 6.0 to 8.0, fungi from 4.0 to 5.0, and actinomycetes from 7.0 to 7.5. Legume bacteria are smaller in number and fail to function in strongly acid soils. Calcium generally favors growth.

Aeration. Air contains gases utilized by soil organisms. Oxygen is needed for respiration, and large amounts are used by the nitrogen-fixing bacteria, fungi, and actinomycetes which oxidize or break down organic matter. Carbon dioxide is used in the process of photosynthesis by some algae and as a source of carbon by some bacteria. Lighter soils have too much aeration, and heavy soils suffer from lack of aeration.

Light. Bacteria and similar organisms will be injured by direct sunlight. Since algae may possess chlorophyll, they often respond favorably to diffused sunlight, as do fungi.

Soil Management Practices Related to Soil Microorganisms

Several important soil management practices control the activities of soil microorganisms. Some of these practices are needed to encourage their work, while others are designed to aid a growing crop which may be suffering from excess microbe activity. Every farming practice or management practice which affects the soil affects such activity.

Aiding Decomposition of Plant Residues. Stubble mulching, the practice of leaving crop residues on the soil rather than plowing them under, reduces the activity of soil microorganisms. When plant residues are mixed with the soil, farmers often supply nitrogen fertilizer in addition, to feed soil microbes and to hasten their work. When this is done, farm crops can continue their normal growth. A general recommendation is to add 20 pounds of nitrogen fertilizer to each ton of straw. Adding phosphorus and potash mineral fertilizers as well as nitrogen may further hasten the process of decomposition. Unless fertilizers are added to crop residues such as straw, the soil microorganisms will use the available soil nitrogen at the expense of growing crops.

Influence of Lime. Soil acidity has an important effect on nitrogen-fixing bacteria. Lime is ordinarily added to soil to

Figure 4-11: The Effects of Lime and Fertilizer. Microorganisms need a nearly neutral, fertile soil in order to release the maximum amount of nutrients for crop growth.

correct acidity. A near-neutral soil generally provides the most favorable condition for soil microbe activity. Lawns or growing farm crops are often greener after lime application, because the lime has stimulated soil organisms to release soil nutrients. Plant decay is slower under acid soil conditions.

Summer Fallow. Farmers in the Great Plains states of Montana, Kansas, Colorado, Wyoming, Nebraska, Oklahoma, Texas, and the Dakotas, under limited rainfall and moisture conditions, practice *summer fallow.* Crops are grown every other year, usually in strips. The ground is cultivated periodically to keep down growth of weeds and other vegetation.

Summer fallow serves *to store nitrogen* as well as *soil moisture.* Moisture is conserved because it is not being used by growing plants. Nitrogen is being stored by the work of nitrifying bacteria.

Moisture and temperature conditions being favorable for bacterial development, summer-fallow soils may store up to 200 pounds or more of available nitrogen per acre in a single season. In addition, soil tilth is improved somewhat by increased activities of soil microbes during summer-fallow months.

Grassland Farming and Soil Microorganisms. Keeping grass and legumes in fields periodically has long been recognized as a soil-building practice. Grassland farming provides abundant supplies of organic matter for soil microorganisms. Thus, soil tilth is improved, nitrogen is added to the soil, and more food elements are made available for farm crops.

Higher Forms of Animal Life in the Soil

In addition to the millions of soil organisms not visible to the human eye, soils may contain large numbers of insects, worms, and rodents. These may be either helpful or harmful.

Slugs and Snails. Slugs and snails are animals which feed on the surface rather than underground. They thrive best under shady and damp conditions. They feed on fallen leaves and old grass, but will attack growing plants when other food is scarce. Their excretions are broken down by bacteria and fungi. Slugs are apt to be a pest in gardens or in potato and beet fields.

Arthropods. Arthropods are insects with jointed bodies and

Figure 4-12: Snail and Slug. Snails, with external shells, and slugs, with internal shells, are molluscs that live on decaying organic matter but sometimes attack valuable crops.

limbs. Examples are ants, centipedes, mites, wood lice, and springtails. Most feed on decaying organic matter, though some may feed on living plants. Some insects take in soil particles with their food. Beetle larvae are important soil burrowers. Wire worms, which attack living plants, are the larvae of beetles.

Nematodes or Eelworms. Nematodes or eelworms are generally abundant in soils. They are 0.5 to 11 millimeters long and about 40 to 50 times as long as they are broad. Nematodes are classified into three groups:

1. A group that lives on plants (usually specific types of host plants).
2. A group that live on other nematodes and other forms of soil life, such as bacteria and protozoa.
3. A group that lives on decaying organic matter, including dead bacteria and protozoa.

The first group of nematodes is of great concern to farmers because of its attacks on crops. Besides puncturing roots, these nematodes prepare for the entrance of other parasites.

Earthworms. In some soils, angleworms or earthworms are important soil builders and conditioners. They are most numerous in loamy soils and are most active in the spring and fall. They like fairly warm and moist soils, and thrive in soil areas rich in organic matter. They prefer sod and clumps of manure. They

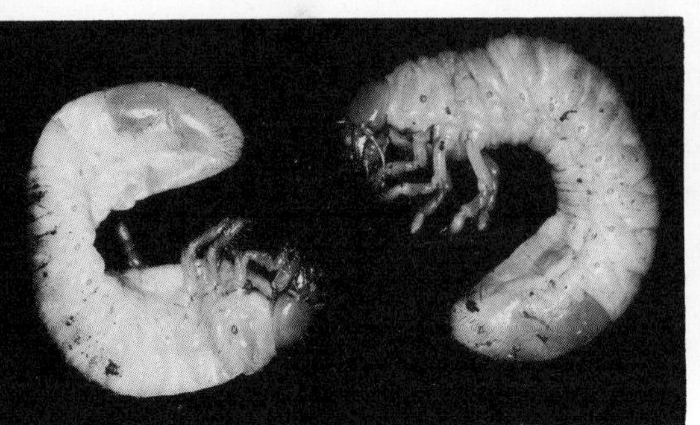

Figure 4-13: White Grubs and Wire Worm. These soil pests are deadly to many valuable crops.

vary in number, size, and weight with the favorableness of the growing conditions.

Studies have shown that certain soils may have millions of earthworms per acre. Under favorable conditions these earthworms will weigh up to 500 pounds per acre.

Some earthworms pick up dead, surface plant materials and mix them with the soil. They drag litter into their burrows, where it is attacked by bacteria and other soil organisms.

Earthworms do not thrive in very acid soils. Such soils are usually lacking in calcium, which these worms require.

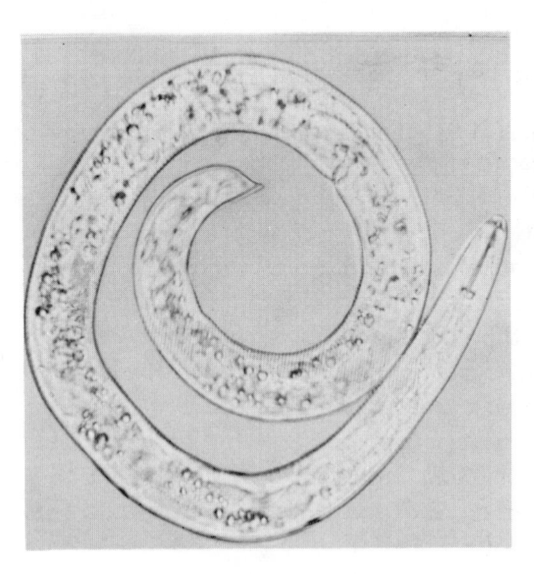

Figure 4-14: Magnified Nematode. Nematodes are another serious soil pest.

Earthworms do extensive burrowing. Their channels permit air to get into the soil. Rain and irrigation water also penetrate the earthworm channels.

An important function of some earthworms is the process by which they take soil and organic matter into their digestive system. In this process, organic matter is ground up with soil particles. This soil mass is mixed with calcium carbonate and excreted by the worm's digestive glands. This excrement is a source of plant food and is high in moisture-holding capacity.

The action of earthworms is most beneficial in forest, sod, and pasture land. Under field conditions, farmers mix the surface soil and organic matter or manures through cultivation. Cultivated fields rich in organic matter can support worm life.

Scientists are not agreed on how much earthworms help to break down plant materials in the soil. Though some earthworms actually take in and cast off soil and plant materials through their digestive tracts, others only secrete substances digested in their body tracts.

Soil Rodents. Larger animals which burrow into the soil include gophers, woodchucks, ground squirrels, and prairie dogs. Activities of these animals sometimes interfere with farming operations, but their effect on soil building and conditioning should not be overlooked. Their activities include soil pulverization, granulation, and mixing. Their burrows permit moisture to penetrate deeper into the soil.

Higher Forms of Plant Life in the Soil: Roots

In Chapter 6, "How Plants Grow," you will learn how roots affect the growth of crops. In the present chapter, the concern is with roots as a part of soil. Roots affect soil during growth and decay.

Root Growth and Soil. Roots in growth exert great force. Probably you have seen concrete sidewalks lifted and cracked by growing roots. In much the same way, roots deep in the ground split and pulverize rock and soil parent materials. They mix the soil and loosen it. They also hold surface soil in place, preventing erosion.

Figure 4-15: Soil-building Roots. The rich, absorbent, well-tilthed soil at left is the creation of the thick root growth that permeates it. The compact, infertile soil at right shows the result of many years of open cropping.

Root Function and Soil. By feeding plants, roots raise soil water and minerals from the subsoil into the topsoil, where some of this water and these minerals will eventually become available to soil bacteria when the roots decay. Roots also move many plant products, such as protein and glucose, from the stems and leaves where they are manufactured into the soil. These materials, too, will eventually be converted into organic matter.

Root Decay and Soil. You have already learned that decaying roots are the most important source of soil organic matter. Decaying roots also add minerals, air, and water to soil. They provide minute channels for the penetration of air and water from the surface, and loosen soil to assist new penetration by growing roots. They improve soil texture and tilth and help prevent erosion, though not as well as live roots.

Summary

Few farmers are aware of the important role played by plant and animal life in the soil.

A handful of soil may contain as many living organisms as

there are people in the world. Bacteria are the most numerous and probably the most important of soil organisms. Other soil microorganisms are actinomycetes, fungi, algae, and protozoa. Soil microorganisms help the farmer by causing the decay of plant residues, thus releasing carbon dioxide, ammonia, and minerals essential to crop growth. They also release essential mineral elements from the soil, making these available for plant use. Some bacteria take free nitrogen from the soil air and convert it into cell material by a process called *nitrogen fixation.* This nitrogen becomes available to plants when the bacteria die. Other bacteria fix nitrogen while living in *nodules* on the roots of legumes. When legumes are grown for the first time, the soil should be inoculated with these bacteria. Under favorable conditions, nitrogen-fixing bacteria attached to legumes can produce 50 to 100 pounds of nitrogen per acre during a year. Some soil microorganisms contribute valuable antibiotic properties to the soil. The by-products of soil microorganisms are thought to improve soil tilth.

Harmful effects of soil microorganisms include bacterial infections and competition for plant nutrients, especially the condition called *nitrogen tie-up.*

Factors that regulate the growth and activity of soil microbes are (1) food, (2) temperature, (3) moisture, (4) acidity, (5) aeration, and (6) light.

Soil management practices are needed to encourage the work of soil microorganisms and to help crops suffering from excess microbe activity. These practices include adding fertilizer to crop residues, liming soils, summer fallowing, and grassland farming.

In addition to microorganisms, soils may contain large numbers of insects, worms, and rodents. These may be helpful or harmful. Included are slugs and snails, arthropods, nematodes or eelworms, earthworms, and burrowing rodents such as gophers.

Roots are the principal higher form of plant life in the soil. Roots are valuable because they help form soil, prevent erosion, raise soil water and nutrients from the subsoil, move plant products such as glucose into the soil, and contribute organic matter, minerals, air, and water to the soil when they decay. They improve soil texture, fertility, and tilth.

Study Questions

1. How well do farmers understand the importance of soil organisms?
2. What kinds of microorganisms exist in the soil?
3. How do these soil microorganisms live?
4. How do bacteria make use of free nitrogen from the air?
5. What are the approved practices of nitrogen bacteria inoculation?
6. How do soil microorganisms release mineral elements for farm crops?
7. What are the antibiotic properties of soil microorganisms?
8. How important are soil microorganisms in bringing about decay of plant materials in the soil?
9. How do soil microorganisms improve soil tilth?
10. How do plant roots improve the soil?
11. What plant diseases are caused by soil microorganisms?
12. How do soil microorganisms compete with farm crops for soil nutrients?
13. What is meant by nitrogen "tie-up"?
14. What conditions are favorable to the growth of soil microorganisms?
15. What are some soil management practices which aid the work of soil microorganisms?

Class Activities

1. Discuss what farm practices of the community are beneficial to soil organisms. Which are harmful?
2. Make a count of the earthworms and insects in a cubic foot of forest or garden soil and estimate their weight and numbers per acre.
3. Visit fields where the effects of soil microorganisms can be studied; observe soil tilth and plant residue decay.
4. Observe the greening up of moist soil, as in a potted glass container, showing chlorophyll activity of algae.
5. Determine the percentage that plant roots represent of the soil weight of a six-inch slice of surface soil in sod.
6. Dig up legume plants to observe nodules containing nitrogen-fixing bacteria.

Soil Moisture

5

In addition to minerals, organic matter, and plant and animal life, soil includes moisture as one of its essential parts.

You may remember that moisture is important in the processes of soil formation. Water breaks down rock by the action of frost, the force of raindrops, and the wearing of streams. It dissolves and combines with certain chemical elements to form acids which eat away solid mineral materials. It is necessary to the life processes of microorganisms, which break down organic matter. It also encourages the growth of roots, which tend to split and mix rock materials in growth, and add minerals, organic matter, water, and air to soil in decay. Since these processes of soil formation continue all the time, soil water is thus of great value in increasing nutrient, organic, and microorganism content of soil, improving texture, and indirectly improving aeration. Some soil scientists believe that water is more useful than anything else in creating soil.

Water also serves as a solvent for many nutrients in the soil, absorbing chemical elements (including fertilizers) into the *soil solution* and thereby making them available to plants. Water itself is a nutrient. It is necessary to many plant growth processes in other ways, which will be detailed later.

In Chapter 5 we first will consider how water enters the soil, how it moves and is stored, and how it is managed. Then we will introduce the following chapters on the growth of plants by concluding this one with a study of why plants need water and how much water they use.

Figure 5-1: Wasted Water, Wasted Soil, Damaged Crop, Flourishing Weeds. The water that is running off this field—and injuring the soil and the crop in the process—might have been held where it fell. Note the gully that seems to be forming in the center background, the soil-smothered rows below it, and the clumps of thistles taking over the left foreground.

Characteristics and Management of Soil Moisture

Crops use several hundred pounds of water for each pound of dry matter produced. An important function of the soil is to provide the crop with an adequate supply of moisture. The farm or ranch operator must do everything possible to increase this supply under most growing conditions. To do this he must understand how water (1) enters the soil, (2) moves through the soil, and (3) is held in the soil.

How Water Enters the Soil

Water enters the soil through large *pores*—air spaces—under the influence of gravity. An ideal soil takes in water as rapidly as it is likely to rain. If an area rarely receives more than an inch of rainfall per hour, soil does not need to take water faster than this. In areas where the rainfall is more intense, a higher intake rate is desirable.

Soil Types and Water Entry. Sandy soils have a high proportion of coarse pores and allow water to pass readily into them. However, the large pores between coarse particles are frequently blocked by finer particles, so that some sandy soils take moisture more slowly than would be expected.

Medium-textured or loamy soils have highly desirable moisture entry capabilities. The individual particles in loamy soils are usually grouped together into clusters or aggregates. These aggregates have fairly large pores between them, so they allow water to enter readily.

Fine-textured or clayey soils are sometimes well aggregated enough to take water rapidly. However, they are less likely to be well aggregated than medium-textured soils. In addition, clays swell when they become wet, causing enough pressure in the wetted zone to reduce the size of larger pores. Many clays crack when they dry. When it rains, water runs rapidly into the cracks until these swell shut. Because such easily cracking soils do not have good structure, once the cracks swell closed further water intake is slow.

How Soil Water Moves and Is Stored

The pores by means of which water enters the soil also partially control its movement within the soil and its storage there. Another

Figure 5-2: Soil Pores. Represented in the enlarged diagram above by white spaces, soil pores provide passages for roots, air, and water. The block at right shows the same amount of soil if compacted.

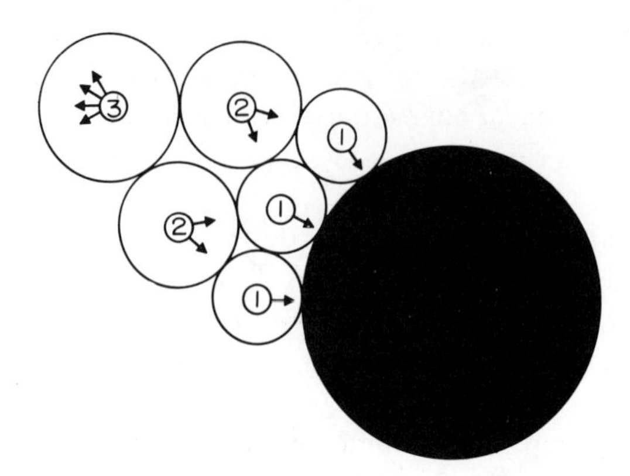

Figure 5-3: Adhesion and Cohesion. Adhesion is the magnetic force that attracts molecules of water (1) to unlike molecules of soil (disk). Cohesion is the force that attracts molecules of water (2) to one another. Water molecules (3) at the outside move away, where water layers are thinner.

very important factor, however, is the characteristic behavior of soil water in relation to soil particles.

Adhesion and Cohesion. Moisture is attracted to soil particles much as steel is attracted to a magnet. Moisture held closest to the soil particle is held with the greatest force, so tightly that it is unavailable to plants. As the layer of water surrounding the soil particle gets thicker, the water is held with less force and plants can remove the excess for their use. As the water layer gets still thicker, it is held so weakly that gravity pulls it down through the soil.

Much of the behavior of water in soils, then, is explained on the basis of *adhesion* and *cohesion. Adhesion* is the attraction of *molecules* of one substance for molecules of another. (Molecules are the smallest portions of an element or compound retaining chemical identity.) Water molecules are strongly attracted to soil minerals and organic materials due to adhesion. Water molecules are also strongly attracted to each other, and this force is know as *cohesion.*

Thus, water molecules are attracted to soil particles directly by adhesion, and indirectly by cohesion. This teaming-up of forces of attraction builds layers of water molecules on the surface of soil particles. Though gravity moves water only downward, the forces between soil particles and water molecules act without regard to direction.

As a water layer on a soil particle becomes thicker, the water near the outside of the layer is less strongly attracted to the soil particle. Thus, there is a tendency for water to move from areas were the water films are thick to where they are thin. This causes water to move into a dry soil. When roots remove water from a portion of the soil, the thickness of the water layer is decreased and water moves in from the surrounding soil to the area that is being dried by roots. Water moves rapidly when moisture films are thick, but as the films decrease in thickness the rate of movement slows. As this slowing takes place, water does not move rapidly enough over more than a fraction of an inch to provide water to roots. Thus, plant roots must be well distributed through the soil to make maximum use of soil moisture.

Many of the pores between soil particles are so fine that even the water in the center of the pore is strongly held to the walls of the pore. As the pores become larger, the water in the center of the pore is no longer held with a force greater than gravity. Under good drainage conditions, gravity pulls the water out of large pores and they are filled with air.

Classification of Soil Water. As we have seen, the force with which water is held to a soil particle increases progressively from the outside of the film to the inside layer of molecules held by adhesion to the soil particles. On this basis, scientists have classified soil water as (1) hygroscopic, (2) capillary, and (3) gravitational.

Hygroscopic Water. Some moisture is held so strongly to the

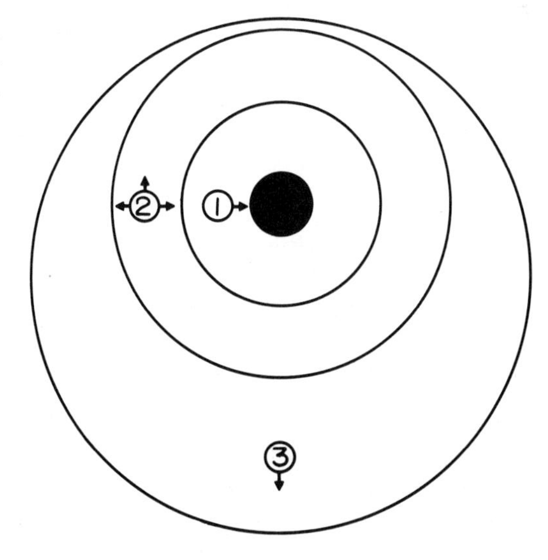

Figure 5-4: Hygroscopic, Capillary, and Gravitational Water. Hygroscopic water (1) is held so strongly to a soil particle (disk) by adhesion that it is not available for plant use. Capillary water (2) is held in the soil by cohesive forces greater than gravity, and moves either upward or sideways from wet places to drier places. Gravitational water (3) moves down through the soil to the water table.

soil particles that it is not available for plant use. This portion of the soil water is known as *hygroscopic water*.

Capillary Water. Capillary water is that portion of the soil moisture which is held by cohesive forces greater than gravity, and yet is available for the growth of plants. It thus lies in a film a little farther from the soil particles than hygroscopic water. While this portion of the soil moisture is held to the soil particles with moderate force, at least the outer portions of the film move fairly readily, either upward or sideways, when differences in cohesive forces occur in different portions of the soil. This type of moisture flow is called *capillary movement* or *capillarity*.

Farmers are concerned principally with that portion of the soil moisture which is available for crops. Capillary water thus is comparable with *available water*. However, the latter term implies, as *capillary water* does not, degrees of maximum and minimum moisture content between either end of the "available" range. These limits cannot be defined sharply but they are put to very practical use.

The moisture content of a recently wetted soil, after gravity has acted long enough to drain out most of the gravitational water (discussed below), is said to be at *field capacity*. Thus at field capacity the soil contains the maximum of available water, the greatest amount of water it can hold against gravity.

Wilting occurs when roots cannot take in water and move it to the upper portions of the plant as fast as water is being lost from the plant. The amount of a soil's moisture content at which plants can no longer recover from wilting is called the *wilting point*. When plants are exposed to a hot dry atmosphere, they frequently wilt even when the soil is wetter than the wilting point. However, when conditions become less rigorous, as during the night, the plants regain "turgor."

"Available water," then, is the amount of water a soil can hold between field capacity and the wilting point. The more available water that a soil can hold in the root zone, the less frequently the soil needs to be wetted by rain or irrigation to meet plant needs.

Gravitational Water. When water films become so thick that attraction forces can no longer hold moisture against the force of gravity, water starts to drain down through the soil. This water is known as *gravitational water*. As long as drainage is satisfactory in loams and clays, gravity draws excess moisture out of the large pores and allows these pores to fill with air. Under poor drainage conditions, gravitational water cannot drain off, and aeration becomes a problem. Sandy or coarse soils do not have adequate moisture storage capacity to prevent excessive drainage, and water which could be used during drought is lost shortly after rains.

Gravitational water moving down through the soil carries dissolved nutrients with it. Losses of nitrogen, potassium, and calcium or lime are especially serious in many soils. This process of *leaching* may be beneficial, however, in correcting alkali conditions.

Gravitational water eventually reaches the *water table*, the level beneath which the soil is saturated with water. This level varies greatly in different regions, tending to be deepest in dry areas and closer to the surface where rainfall is plentiful; marshes result when the water table is just below the surface of the ground. If the water table is not too low, it tends to correct surface dryness by capillary movement. Very low water tables require wells, pumps, and surface irrigation if the farmer is to use the water there. Very high water tables make drainage necessary for the successful growth of most crops.

Figure 5-5: Soil Auger. The soil auger provides a quick, easy way to determine the amount and depth of moisture in a field.

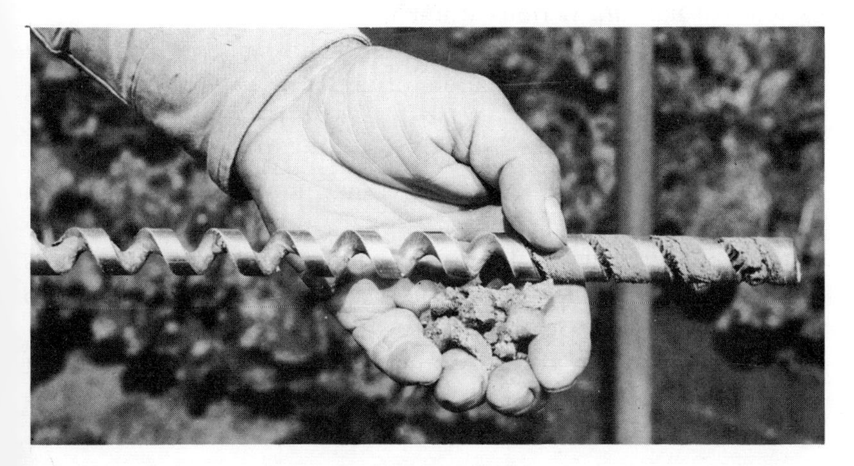

Amount of Water Held in Soils. Because, as we have seen, water is held in soil by forces attracting it to the surface of soil particles, the greater the surface area of the soil particles, the more water the soil will hold. As a material is broken into finer particles, more surface is exposed. The importance of size of particle in relation to surface area is readily apparent when we consider the surface area of a closed book in comparison with the surface area of all its pages. Thus a soil which consists mostly of coarse particles has a low ability to hold water, whereas finer soils hold more water.

In general, then, sandy soils have low water-holding capacity in comparison with clayey soils. Very sandy soils will frequently store one-half inch or less of rain for each foot of depth. In such a soil, two inches or less of water would be stored in a four-foot depth of soil. Loamy soils frequently hold one to two inches of available water for each foot of depth, or four to eight inches of water in a four-foot depth. Clay soils may hold up to three inches of available water for each foot of depth. It should be remembered in this connection, however, that clay soils are often too poorly aggregated to permit large amounts of water to enter them under many normal rainfall conditions. Thus, loamy soils often store larger amounts of water than clay soils with superior moisture storage capacity, because loamy soils permit more water to enter.

The amount of water a soil will hold, then, depends on the size of its particles, the depth of the soil, and the extent to which its relative *porosity* permits water to enter.

Determining Soil Moisture Content. The moisture content of soils can be determined in laboratory tests by weighing a sample of soil before and after moisture has been removed in an oven. The sample is heated to temperatures of 105 to 110 degrees centigrade. The moisture content is commonly expressed as a percentage of the dry weight.

Scientists also have developed methods of measuring the moisture content of soils under field conditions. Among these is the use of a small porous ceramic (baked clay) cup filled with water. The tendency for the soil to extract water from the cup is in proportion to the moisture content of the soil, and a suitable gauge will measure this. A second method is to measure the

Figure 5-6: Instruments for Measuring Soil Moisture. At left, soil scientist George J. Bouyoucos of Michigan State University demonstrates his device for measuring moisture by the amount of electricity conducted by soil-moistened plaster-of-Paris blocks. At right is a meter that uses radioactivity to determine the amount of soil moisture.

electrical conductivity between two electrodes buried in a porous plaster-of-Paris block. The conductivity is proportional to the moisture content of the block, which is equal to that of the soil. A third method is based on the heat conductivity of the soil, heat conductivity also depending on the moisture content. The absorption of radioactive materials by soils also being proportional to moisture content, this fourth method of determining moisture content is sometimes used.

Management of Soil Water

The rate of growth of plants is generally in direct proportion to the available moisture, though too much water may the harmful. The soil moisture supply depends on (1) the amount of snow, rainfall, and irrigation and its distribution in relation to the growing season, (2) the rate at which the soil permits water entry, (3) water storage capacity of the soil, (4) evaporation rate, (5) the height of the water table, (6) the rate at which water can move within the soil and rise from the water table by capillarity, and (7) the amount and nature of vegetative covering.

Increasing Available Water. Under most farming conditions, an increase in the amount of available water is desirable. The following is a summary of ways to accomplish this, many of which are discussed in more detail later in this book.

Precipitation. Precipitation is the fall of rain or snow. The question of whether precipitation can be increased by such artificial methods as seeding clouds with dry ice is as yet unanswered. Even if artificial rainmaking should prove to be reliable, however, all farmers will continue to find it beneficial to take water conservation measures.

Snow may be prevented from blowing away with the help of fall plowing, mulches, and windbreaks. It is probable that terraces also contribute to snow conservation. Many farmers find that fields covered with snow during the winter are ready for spring plowing about as soon as bare fields, because snow-covered fields generally freeze less deeply.

Runoff of rainwater is reduced by contour plowing, contour planting, strip cropping, crop rotation, terracing, and the use of grassland farming, mulches, and cover crops. Careful crop selection may also be of value, close-growing crops tending to hold more water on the field than row crops such as corn. Growing crops will lessen the force and angle of wind-driven rain, thus lessening the danger of erosion.

Irrigation. This topic is treated fully in Chapter 16. Many farmers in regions of relatively high rainfall are installing

Figure 5-7: Stubble and Snow. The standing wheat stubble at left has held nine inches of valuable snow on the field, where its moisture will be used by the summer's crop.

Figure 5-8: Water Conservation at Its Best. Terracing, contour plowing, and strip cropping are combined on this Texas farm to hold water on the field where it is needed.

portable irrigation systems to take advantage of the fact that irrigation can be timed exactly, as natural rainfall at present can not, to supply water when it is most needed during the growing season.

Entry. Entry rate is determined by the size and number of soil pores, a few large pores taking water faster than many small pores. Plowing, which loosens soil, will temporarily increase entry rate, but its benefits decrease as the soil settles back into its original state. More permanent improvement may be supplied with green manures, which help aggregate sandy soil and loosen clay. Correction of alkaline conditions is another method of improving water intake.

It is important to remember in connection with entry rates that the devices for holding water on the soil (described above under "Precipitation") will compensate for low soil porosity. A clay soil which has been plowed on the contour, for instance, will have time to absorb a great deal of water. Since the water

Figure 5-9: Plowing and Entry Rate. The tops of the stakes across these contoured pasture grooves show the depths of moisture penetration as measured by a soil auger.

storage capacity of clay is very high, the amount of water stored may be greater than that stored by loams and sands with superior water entry rates.

Storage Capacity. As we have seen, the water storage capacity of a soil depends upon the size of its particles and upon its depth. Storage capacity of sandy soil can be improved by the use of green manures and animal manures, which will add fine, water-retaining particles of organic matter. Correction of acidity and alkalinity is another beneficial device. To the extent that the improvement of storage capacity of loams is necessary, this also may be accomplished by use of green manures and animal manures. The problem with clays, of course, is not storage capacity but entry rate.

Evaporation. High temperature, low humidity (water content of air), and wind encourage soil drying. For some time after a rain, the surface soil stays moist because water moves to the surface almost as fast as it evaporates. This tends to thin the water films throughout the soil mass. As the films become thinner, movement cannot take place as readily. Flow to the surface can no longer take place as fast as evaporation, and the soil surface dries.

Various attempts have been made in the past to reduce the amount of water lost by evaporation. The method best known in areas of limited rainfall was the *dust mulch* or *dry mulch,* the establishment of a loose, dry surface layer of cultivated soil. However, the water lost in establishing the mulch is now thought to be about equal to that lost without it. Today, farmers are experimenting with mulches of crop residues and/or manure, hay or straw, and even—between rows of crops—strips of plastic or metal foil.

Figure 5-10: Trashy Ground Cover. Crop residues left on the soil surface help decrease evaporation as well as wind erosion.

It is generally agreed that the drying effects of wind can be limited to a considerable extent in both summer and winter with the help of evergreen shelter belts, strip cropping, crop rotation, grassland farming, plowing at right angles to the direction of the prevailing wind, and in some cases by terraces. Ground covers also give some wind protection. Since wind protection permits humidity to rise within the protected area, such protection tends to reduce the drying action of low humidity as well. You can demonstrate the resistance of enclosed air to drying by putting a wet milk bottle in front of a fan and contrasting the length of time it takes the inner sides of the bottle to dry with the length of time it takes the outer sides to dry.

Water Table. General conservation measures, discussed in Chapters 17 through 19, can substantially raise the water table. This is important because where the water table is relatively high it can be tapped by roots, either directly or through capillary action, while where it is relatively deep it can serve as a reservoir for irrigation. Some farmers irrigate plants by raising water tables rather than by applying water to the surface of the soil. Water raised from the water table either naturally or artificially is likely to have a high mineral content and encourage salt accumulation.

Capillary Rate. Capillary rate is controlled by soil structure and texture. Like entry rate, this may be improved as necessary in the different soils by variously plowing, using green manures and animal manures, and correcting acid and alkaline conditions.

Vegetative Covering. Both the amount and the nature of vegetative covering affect the level of available soil moisture.

In respect to the *amount* of vegetative covering, in areas of relatively high precipitation, such as the Eastern states, vegetation is known to increase available soil water. Leaves and stems tend to hold rainwater and snow on the soil until entry can take place. In areas of limited precipitation, however, the additional water which leaves and stems might help enter does not compensate for the rate at which plants absorb soil water and release it into the air. Plants also intercept a certain amount of water which remains on the plants until it evaporates, so that this water never reaches the soil surface. Because the water that enters the soil in such regions is usually lost from the surface,

Figure 5-11: Summer Fallow in North Dakota. The fallow strips have been laid out at right angles to the prevailing winds in order to decrease wind erosion.

either by direct evaporation or through plants, little or no leaching takes place and high fertility is a desirable result. To take advantage of this fertility, farmers practice *summer fallowing*. This is a system of farming by which fields are kept bare of crops and weeds one year, to store water for the next.

In land under summer fallow, relatively little water is lost by evaporation from depths below four inches, and none, of course, is used and given off by plants. Though the loss of water by surface evaporation continues to be more serious than in humid areas, because light rains may be evaporated soon after they fall, heavier rains penetrate deep enough to resist evaporation.

Western Great Plains farmers and ranchers know that crops of grain on some lands would be impossible without this system, which conserves about 20 per cent of the moisture that falls during the fallow period. In areas with ten to 20 inches of annual precipitation, this is enough to help assure a grain crop the following year.

In respect to the *nature* of vegetative covering, farmers should remember that different crops, and different varieties of these crops, have widely varying water requirements. These variations are discussed in somewhat more detail at the end of this chapter.

Another fact to remember is that weeds deprive more desirable plants of water. Weeding is thus an important means of soil moisture conservation.

Decreasing Soil Water. Land drainage is treated in detail in Chapter 15.

Soil Moisture and Crop Growth

Soil water is essential for the growth of both tops and roots of plants. Water makes up 75 to 95 per cent of plant tissues. It serves (1) as a plant nutrient, (2) as a carrier of plant food, and (3) as a part of each plant cell.

Water As a Plant Nutrient. Water serves as a plant food or nutrient along with other raw materials. Plants use water and carbon dioxide to form sugars and more complex carbohydrates.

Figure 5-12: Diffusion and Osmosis. The diagram at left represents the equal distribution of molecules in a solution as a result of diffusion. The diagram at right shows how small molecules diffused in a soil solution (1) pass through the small openings of a semi-permeable membrane (2) into a plant cell (3). Since the large molecules within the cell cannot pass through the membrane, the crowding of molecules within the cell creates osmotic pressure.

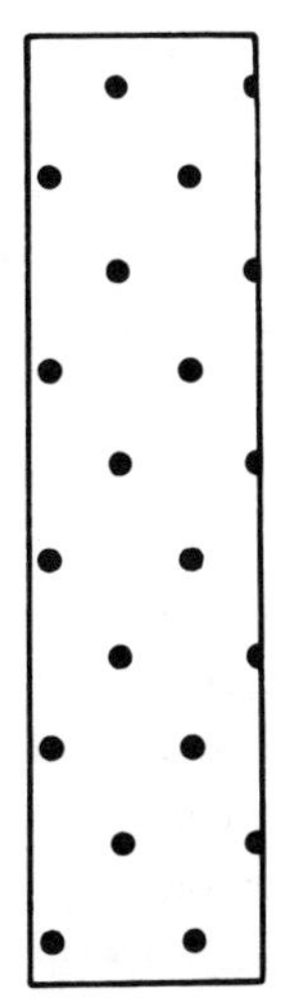

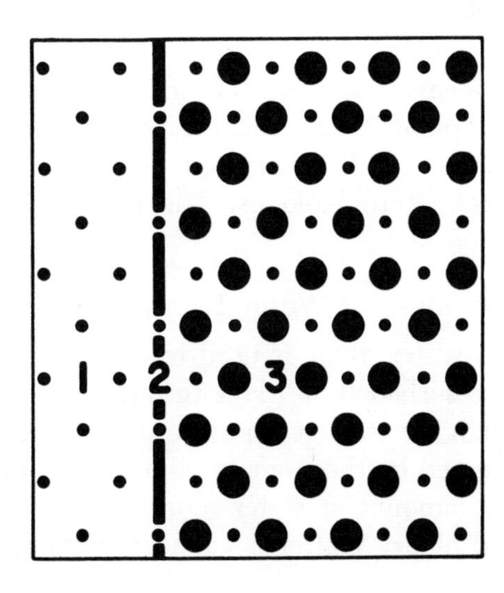

Water As a Plant Food Carrier. A vast amount of water moves throughout the plant. Essential nutrients from the soil are absorbed into the plant through water solutions. Within both the soil and the plant, these nutrients move through water solutions by *diffusion*. Diffusion is the process by which molecules or ions in a solution move about freely until all parts of the solution have an equal concentration. You can demonstrate diffusion by dropping a lump of sugar into a glass of water; the sugar will dissolve and its molecules will disperse throughout the water. As nutrients are absorbed from a solution by roots, additional nutrients move by diffusion into the area surrounding the roots.

Osmosis. Osmosis is a form of diffusion through a *semi-permeable membrane*, such as the wall of a plant cell. In respect to plant growth, osmosis differs in its effects from those of simple diffussion because the molecules of the soil solution are frequently smaller than the molecules in the plant cell. Small molecules pass more readily through the openings in the plant cell membrane than large molecules. Thus, the number of small molecules within the plant cell will increase without a corresponding loss of large molecules. Since this pushes the large molecules farther apart, a pressure, called *osmotic pressure*, is built up inside the plant cell. Differences in osmotic pressure account for much of the movement of water and nutrients throughout the plant.

Water As a Plant Constituent. Water is the major constituent of a plant cell and consequently of the total plant. Protoplasm, the complex living substance of cells, includes water, protein, fats, sugars, and salts. The rigidity of the plant cell, and of the whole plant in the case of herbaceous plants, is due to water pressure on cell walls. Cells in effect are blown up with water like miniature balloons. The plant wilts when water pressure gets too low.

The Amount of Water Transpired by Farm Crops

Soil water, then, is used by plants in the following ways: (1) It is absorbed by a plant through its root system. (2) It passes through the stem system into the cell bodies of the leaves. (3) It is released by the stomata (pores) of leaves by *transpiration*.

The amount of water required by various crops to produce a pound of dry matter is called the *transpiration ratio*. In careful

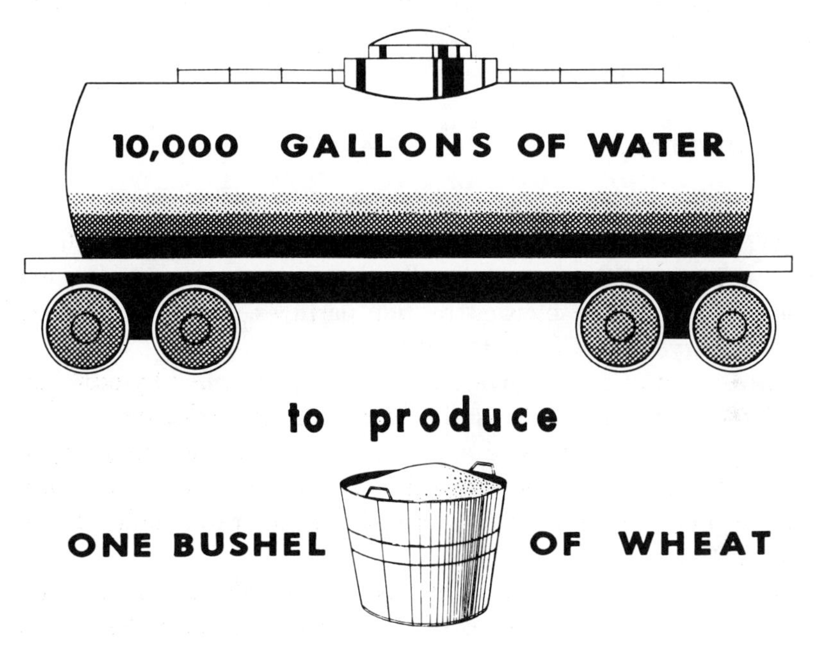

Figure 5-13: The Reason for Moisture Conservation. The water that a farmer can hold on the field for use by his crops increases his profit.

Colorado experiments, conducted in a dry climate, the transpiration ratio was approximately 300 pounds of water for millet and sorghum, 350 pounds for corn, 500 pounds for wheat, 630 pounds for potatoes, 800 pounds for clover, and 860 pounds for grass or alfalfa. These ratios vary from year to year and area to area.

Measuring Water Use by Crops. Studies indicate that the amount of water used by a crop is related to the amount of water evaporated from an open tank. Research workers thus can relate the previous rainfall, the water storage capacity of the soil, and the nature of the crop to evaporation from standard tanks. In this way they are able to predict when a crop should be irrigated.

In general, such research has established that plants may use as much water during one hot dry day as they use in a week of cool humid weather, since temperature and humidity govern the amount of water used by plants. Similarly, the amount of water transpired by crops in arid or hot regions is greater than in humid or cool regions.

Summary

Moisture is an essential part of soil.

Water enters soil under the influence of gravity through large pores. In general, sandy soils allow water to enter rapidly, but their pores may be blocked by fine particles. Loams have good water entry rates. Clays often do not.

Movement and storage of soil water are controlled by adhesion and cohesion. *Adhesion* is the attraction of molecules of one substance for molecules of another. *Cohesion* is the attraction of molecules of a substance for other molecules of the same substance. Both forces tend to attract layers of water molecules to soil particles.

Soil water is classified as hygroscopic, capillary, and gravitational. *Hygroscopic water* is held so strongly to particles that it is not available to plants. *Capillary water* also is held against gravity by particles, but it is available to plants. It moves without regard to gravity by capillarity. Capillary water is comparable with *available water,* but the latter term includes all degrees of soil wetness between field capacity (the maximum amount of water a soil will hold against gravity) and the wilting point (the level of moisture content in a soil at which plants can no longer recover from wilting). *Gravitational water* is that part of soil water that moves down through the soil to the water table, the level beneath which the ground is saturated.

Clay stores the most water, sand the least. Moisture content of soil may be determined in both the laboratory and the field.

The soil moisture supply depends on (1) the amount of snow, rainfall, and irrigation and its distribution in relation to the growing season, (2) the rate at which the soil permits water to enter, (3) the water storage capacity of the soil, (4) evaporation, (5) the height of the water table, (6) the rate at which water can move within the soil and rise from the water table by capillarity, and (7) the amount and nature of vegetative covering. It is possible to manage these factors to increase soil moisture.

Soil water serves (1) as a plant nutrient, (2) as a carrier of plant food, and (3) as a part of each plant cell. The amount of water required by crops to produce a pound of dry matter is called the *transpiration ratio.* This varies for different crops. The amount of water used by a crop is proportional to the amount of water evaporated from standard tanks.

Study Questions

1. How does water enter the soil, and what are the characteristic entry rates of the three major soil types?
2. What are adhesion and cohesion?
3. What is capillarity?
4. What are the three kinds of soil water, fully defined?
5. What are field capacity, the wilting point, and available water?
6. What are the water storage capacities of the three major soil types?
7. How is soil moisture content determined?
8. What seven factors control the soil moisture supply?
9. How may each of these seven factors be managed to increase the amount of soil water?
10. In what three ways is soil water essential to crop growth?
11. What are diffusion and osmosis?
12. What is the transpiration ratio, and how does it differ among various crops?
13. How is water use by crops measured?

Class Activities

1. Demonstrate the upward capillary movement of water by use of different-sized capillary tubes placed in a basin of water. The water will rise in the tubes above the surface level of the outside water. The attraction of the glass for the water, called *adhesion,* will cause the water to rise around the sides of the tube. The *cohesion* force of the water molecules for one another will cause the water column to rise as it is further assisted by adhesion

forces. Note the greater adhesion tension in the smaller capillary tube, as shown by the greater water curvature at the top of the water column.

2. Get two or more five-gallon soil containers with bottom drains. Demonstrate how fast gravitational water moves through soils of different tilth and texture.
3. Observe the capillary moisture action in a cupful of dry loamy soil placed in a conical pile on a pie tin to which about a third of a cup of water is added.
4. Take field trips to examine the moisture-holding capacity of soils of different textures. Press the soil into balls to observe water-holding properties.
5. Recall community practices for moisture conservation, such as summer fallow and weed eradication.
6. Conduct laboratory demonstrations on the effect of mulches on the evaporation of water. Fill three containers of equal size with silt loam within 1½ inches of the top. Add enough water to each container to bring the soil moisture up to field capacity. Add one inch of chopped straw to one container and one inch of dry silt loam to a second container. Keep the third container as a check. Weigh each of the containers each day and record the loss in weight due to evaporation.

How Plants Grow

The life processes of plants have startling similarities to those of human beings and animals. Plants and animals share birth, youth, adult life, reproduction, and death.

Only by knowing how plants grow can a farmer intelligently aid their growth.

The following vital life processes that take place in plants will be discussed in this chapter: (1) photosynthesis, (2) transpiration, (3) respiration, (4) digestion, (5) assimilation, (6) growth, and (7) reproduction. Other discussions will deal with (1) the structure of the plant cell, (2) the function and extent of the plant root system, and (3) the functions of stems.

Plant Cells

Life processes of plants are carried on in millions of tiny living cells. Cells vary in shape, size, thickness of walls, chemical composition, and the work they perform. Plant cells can absorb nutrients—that is, they can digest food materials and build them into *protoplasm,* which is the living substance of the cell body. The life processes of plant cells are activated by a group of complex proteins called *enzymes.*

The three major parts of a cell are (1) the cell wall, (2) the protoplasm, and (3) the vacuole.

The Cell Wall. Cells of higher plants are tiny compartments enclosed in a tough, elastic *cell wall.*

The Protoplasm. Occupying the interior of the cell is a living substance called *protoplasm.* The protoplasm of the cell contains a number of important life-giving particles. One of these is the *nucleus.* Hereditary factors which influence the development of plants reside in the nucleus. All the protoplasm outside of the nucleus is known as *cytoplasm.* It is a gel-like substance. Imbedded in cytoplasm are numerous bodies known as *plastids.* They are usually centers of certain types of physiological activity, and are commonly classified on the basis of their color into three groups. The bodies of one group, which are colorless, are called *leucoplasts.* The bodies of a second group, which contain green chlorophyll pigments and yellow pigments, are called *chloroplasts.* The bodies of the third group, which contain red or yellow pigments, are called *chromoplasts.* A single plastid may be colorless, green, red, or yellow at different periods of its existence. The red and yellow colors of some fruits and flowers are formed by chromoplasts. Chloroplasts are the factory units of the cell, in which the process of photosynthesis takes place, as described below.

The Vacuole. The bulk of the interior of a mature plant cell is occupied by a single large cavity known as a *vacuole,* which is filled with *cell sap.* The cell sap is composed of water and plant nutrients.

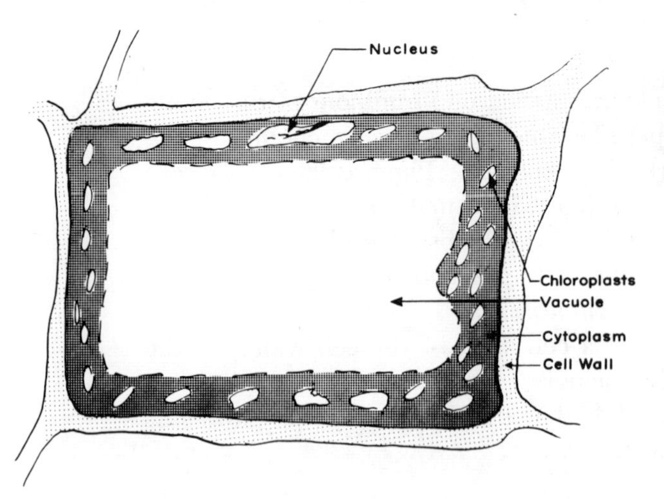

Figure 6-1: The Unit of Life. The chloroplasts shown on this enlarged cross section of a leaf cell are one kind of plastid. Can you name the other two?

Figure 6-2: Light, Air, Water. Abundant light, air, and water permit these healthy bean plants to produce food by photosynthesis.

Life Processes of Plants

Photosynthesis. One of the marvels of nature is *photosynthesis,* which takes place in the chloroplasts of leaves and other green parts of plants. In this process, two common compounds, water and carbon dioxide, are utilized to form sugars and oxygen. The term *photosynthesis* itself describes what happens. *Synthesis* means "making" or "building up" or "putting together" a compound from simpler elements or compounds. In this case, water and carbon dioxide are the materials used. *Photo,* meaning "light," refers to the use of light for energy. *Glucose* is one of the principal sugars thought to be produced.

The glucose molecule contains transformed and stored *light energy.* Plants further transform this energy by converting glucose into other chemical compounds, such as (1) sucrose (beet or cane sugar), (2) cellulose (cell-wall material), (3) fatty substances, or (4) protein, by the addition of nitrogen, sulfur, phosphorus, and other elements.

Photosynthesis is said to be the commonest chemical process on earth. Even coal and oil were first produced, in the form of plant materials, by photosynthesis.

Transpiration. The loss of water in the form of vapor from a living plant is called *transpiration.* Most of the water vapor passes out through tiny openings in the leaf, called *stomata.*

More than 99 per cent of the water absorbed by plants is given off in this way. It has been estimated that the annual transpiration of an acre of corn would flood the land to a depth of 15 inches. The amount of water that plants must absorb to produce a pound of dry matter varies from 250 to 1000 pounds. These figures account for the effectiveness of summer fallowing as a water conservation device.

Respiration. The process by which the plant releases the energy stored by photosynthesis is called *respiration.* Respiration is thought to be controlled and regulated by *enzymes,* which are complex protein substances. The living protoplasm in a single cell breaks down carbohydrates and fats and thus releases energy. The by-products of respiration are usually *carbon dioxide* and *water.* In other processes of respiration some organic acids are formed. In scientific language, respiration in plants is the *oxidation of organic substances,* with the release of energy, within cells.

Many of the life processes of the plant from the root tips to the leaves get their energy from respiration. Plants need energy to manufacture more complex food substances such as proteins, to transport food materials to various parts of the plants, and to absorb minerals. Respiration takes place in all the living cells of the plant.

Digestion. Plants as well as animals digest food. Digestion is carried on in plant cells in which food is stored. During photosynthesis, the plant produces a reserve of sugars which are transformed into starches. As the plant needs more food energy, the starches are changed back into sugars. This is an example of plant digestion. Sugars can be moved from one area of the plant to another. Fats and proteins similarily need to be digested to be broken down for energy and for purposes of *translocation,* movement from one plant part to another. The digestion process is greatly aided by enzymes.

To make the difference between digestion and respiration clear, it might be said that digestion is the process by which plants convert food reserves stored after photosynthesis into forms that permit respiration to take place.

Assimilation. The process by which a plant converts food materials into protoplasm and cell walls is known as *assimilation.*

The plant is able to build up sugars into starches and fats. By the addition of nitrogen, sulfur, phosphorus, and sugar products, the plant produces proteins. From these food substances, along with other essential elements, the cells manufacture protoplasm. The process of assimilation takes place in all cells.

Growth and Reproduction. Anyone working with plants or farm crops has thrilled to watch them grow, mature, produce flowers, seeds, fruits, and an abundant harvest. Students of soils and plant husbandry who know the growth areas of plants can better protect and assist their growth.

The two major processes associated with growth are (1) the vegetative development of roots, stems, and leaves, and (2) the reproductive processes involving the development of flowers, fruits, and seeds. Plant growth consists of an increase in size as well as in weight.

The first phenomenon of growth takes place in the seed. As the seed absorbs moisture, it begins to expand. The seed coat softens. This softening process enables the first tiny plant parts to emerge. The food that plants use for this early growth is contained in the seed itself.

Figure 6-3: Ready for the Seed. Finely pulverized seedbeds make it sure that seeds will be in contact with moist soil.

Once the seedling has broken through the ground, it is greatly aided in its process of growth by photosynthesis.

Plants have three main stages in their above-ground growth. In the first stage, plants make a rather slow start while building up their root system. In the second stage, visible growth is rapid. The third stage is marked by a slowing of vegetative growth but also by the formation of seed, fruits, tubers, or root crops.

Meristems. Plant growth takes place as a result of cell division and elongation. Areas of the plant in which cell division takes place are called the *meristem* regions. The major *growth points* associated with the lengthening-out process are the tips of stems and roots. The second important area of growth is associated with the increase in the diameter or circumference of a stem or root.

Meristem regions are (1) at the tips of branches, (2) near the tips of roots, (3) in the cambium layer of stems, (4) at the bases of internodes on stems, and (5) at the bases of sheaths and leaves, in the case of grasses.

Plant cells increase in size largely from the pressure of the water they absorb, the cell walls stretching much as would a bladder when filled with water. The physical make-up of the cell causes both elongation and an increase in diameter.

Growth in plants is qualitative as well as quantitative. The formation of leaves, flowers, seeds, fruits, roots, and tubers are examples of qualitative growth. This feature of qualitative growth has its beginning in plant cells. Some cells may perform the single function of increasing size. Other cells may develop to produce the different seeds, and the like.

By understanding and appreciating the growth processes in plants, farmers can better aid plant growth and account for disturbances in the normal growth pattern.

The Function of Roots

The root system of a plant performs several vital functions. It absorbs water and mineral salts. It anchors the plant to the soil. Roots also conduct food manufactured in the leaves throughout the root system. In some crops, such as root vegetables, roots store food.

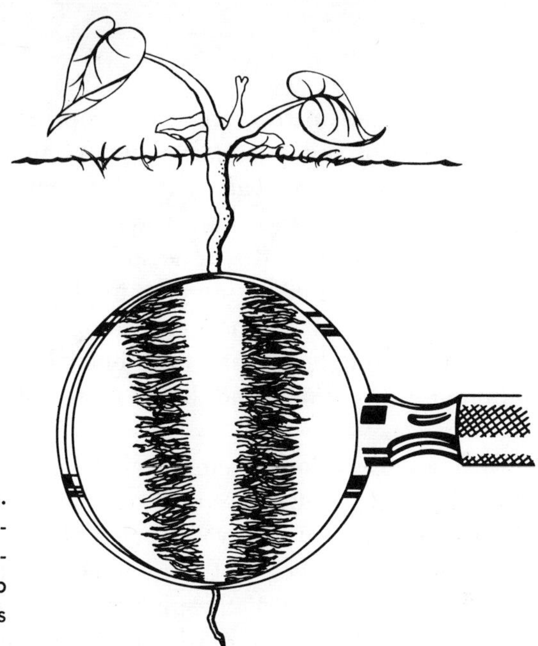

Figure 6-4: Root Hairs. Root hairs, here magnified, are single, elongated cells that absorb water and nutrients from the soil.

How Roots Grow. The growing parts of roots are usually white. These growing root parts absorb water and soil nutrients. Older root parts turn brown in color and are corky in texture.

Root hairs are sent out near the growing tips of roots. They have a slimy surface, and fine soil particles stick to them. These root hairs are believed to be short-lived. Their death and decomposition are sources of food for soil microorganisms.

Root growth is promoted by favorable moisture, food, aeration, and soil tilth. Roots are supplied with food in the form of carbohydrates manufactured in the leafy parts of the plant. When the surface part of plants requires most of the carbohydrates for its own needs, the root growth suffers. This is the case during the production of fruit and seeds. It is partly to promote good root growth that farmers raising fruit often reduce the number of blossoms on trees, bushes, and vines. Similarly, if pasture or range is grazed excessively by cattle, the root growth is diminished.

Depth of root growth is related to the available soil moisture. Grain crops with good moisture conditions have good top and root growth. Roots of some plants, such as alfalfa, may penetrate the soil by as much as 30 feet.

The roots left in the soil when crops are harvested are a valuable source of organic matter and plant food for future crops. Weight of roots will vary with different crops.

Farmers have noted that growing two different crops, such as wheat and barley, in the same soil at the same time will increase the total yield, while no increase in yield is secured from growing different varieties of the same crop. This phenomenon is explained by the facts that different plants require nutrients in different proportions and have root systems favoring different soil zones.

Water Absorption. As much as 99 per cent of the water absorbed by the plant passes up from the stem to the leaves and out through the stomata as water vapor. In most plants, the stomata are in large numbers on the underside of the leaf. Water makes up 75 to 95 per cent of the plant tissue.

The area of contact of the root system with soil and water particles is enormous. The molecules of water must penetrate the cell walls of the root hairs and other thin-walled root cells. The concentration of the cell sap in a root hair and root tip is normally greater than that of the soil solution, so osmosis (explained below) is possible. Once in the root system, the water moves from the very small root branches to larger roots and finally to the stem and upward to the leaves.

Roots only grow in moist soil. Maximum root growth takes place only if the soil is moist at a depth that roots can penetrate. Plants with a deep root system can better withstand temporary periods of drought or low moisture, since they have a larger moisture reservoir to draw upon.

When the intake of water by a plant is less than the rate of transpiration, the plant wilts. To prevent this, if climatic conditions favor rapid loss of water vapor through the leaf openings, the root absorbs larger amounts of water if it is available in the soil.

The movement of water into and out of living cells, you remember, is explained by the process of *osmosis.* Water tends to move through the permeable cell walls from regions of low solution-concentration to regions of high concentration. This tendency is counteracted or brought into balance by the pressure of the cell contents on the elastic cell wall.

Figure 6-5: The Cost of Overgrazing. When pasture is overgrazed, roots disappear along with top growth. Gullied wasteland like this can be the result.

Soil nutrient materials in *ionized* (consisting of electrically charged atoms) form are not similarily absorbed on the basis of osmotic pressure. These can penetrate the cell wall and develop concentrations in the cell far higher than concentrations in the soil solution. The absorption of mineral element ions is essential to the life processes taking place in the plant cell.

A direct relationship has been observed between the extent of the root system of a plant and its above-ground growth. The leaf growth above ground furnishes food for the root system, and the root system in turn furnishes minerals and water for the above-ground leaves and stems. Any disturbance of the root system or the above-ground plant growth affects the other. The top growth and the root system work together as a team. Overgrazing of the leaf growth of pasture and range plants will decrease the growth of the root system. Injury from grazing includes both loss of leaves and destruction of the growth areas of plants.

Absorption of Mineral Elements. Plants may absorb any of the mineral elements in the soil, but are selective in the kind of elements that they absorb in large quantities. Chemical elements absorbed in large amounts are nitrogen, potassium, sulfur, phosphorus, calcium, magnesium, and silica. Elements found in plants in small quantities or mere traces include iron, manganese, boron, zinc, copper, molybdenum, iodine, and selenium.

Green plants get their nitrogen largely from nitrates and ammonia compounds, their phosphorus from phosphates, and their sulfur from sulfates. The principal metallic elements, potassium, calcium, magnesium, and iron, are secured with nitrates, sulfates, and phosphates.

Substances that enter the plant cells of the root system are generally in solution in water. Soil solutions contain gases and mineral salts in addition to minerals.

Extent of Root Systems. A better appreciation of plant culture results from an understanding of the extensive root system of plants. The total weight of the root system may exceed the weight of the top growth, and the number and extent of principal root

Figure 6-6: Top Growth Vs. Root Growth. This photograph only suggests the extent to which the bulk of root growth exceeds that of top growth.

branches may equal or exceed the total number and extent of stem branches. A sugar beet at the end of its first season's growth may have a root system five to six feet deep and lateral branches of equal length.

The total dry weight of roots of mature prairie grasses in temperate subhumid areas runs as high as five tons per acre in the plow area of the soil. The roots of cultivated grasses may run two to three tons per acre for bluegrass and one to two tons per acre for alfalfa and sweet clover.

An appreciation of the extensiveness of root growth is revealed in a study of the roots and root hairs of a single rye plant by H. J. Dittmer at the University of Iowa. First, single rye plants were grown in wooden containers. When the plants were four months old, the soil was washed from the roots and the entire root system was removed intact. One rye plant had 13,815,672 roots, with a surface area of 2554 square feet. Living root hairs on the plant numbered 14,375,568,288. They had a total surface area of 4321 square feet. The total root system surface area was 6875 square feet. The combined length of all its roots was 387 miles. An average of 100,000 new root hairs a day were produced in the plant. The root system of the plant would have had to grow an average of three miles a day to reach this growth. The surface area of the above-ground parts of the plant totaled 51.5 square feet. The surface area of the roots was 130 times greater than the above-ground growth.

Few farmers realize how extensive root systems are. Root systems of crops grown in rows overlap each other much sooner than the top foliage. The root systems of wheat or alfalfa, when grown in wide rows, are much more extensive than those of the same or other crops when planted in the usual manner.

Thus, alfalfa is grown in rows in some dryland areas to permit the plants to develop extensive root systems. Corn crops are at times planted in wide rows with intervening legume and grass crops. These "intercropping" practices are based upon an understanding of the root systems of these complementary crops. In addition to conserving moisture, the planting of legumes and grass along with corn may serve as a soil-building practice. Intercropping in itself, however, may not eliminate competition from root growth. Farmers need to know (1) the depth roots

Figure 6-7: Intercropping. A mixture of grasses and legumes is being seeded between wide rows of corn. How does this practice serve to build the soil?

penetrate, (2) their horizontal spread, and (3) their rate of growth. Such knowledge will enable farmers to regulate tillage, irrigation, and fertilizer practices.

Root Hairs. Root hairs are important nutrient-absorbing and water-absorbing organs. Most abundant near the tips of very young roots, root hairs are single elongated cells. In most plants, the life of a root hair is a few days or weeks.

Balance Between the Top Growth and the Root System. Plants given a chance for normal growth will maintain a balance between root system and top growth. The top growth of the plant manufactures the carbohydrates which supply its own needs and those of the root system. The root system in turn has the task of absorbing water and soil nutrients to supply the total plant's needs. The balance is one of food and water rather than one of size. The top growth may also store a part of its food in the root system. The growth of the common potato tuber is an example of this. Starch materials produced in the leaves are stored in the tubers underground. The same is true of root crops such as sugar beets, carrots, radishes, and turnips.

The balance of roots and top growth may be disturbed by (1)

diseases or insect pests that attack either the roots or the shoots, (2) root pruning from too deep or too close cultivation, (3) breaking of roots during transplanting, (4) pruning of branches, (5) excessive grazing as in the case of pastures and ranges, and (6) harvesting.

Range management experts point out that overgrazing diminishes the extent of the root systems of range plants as well as the top growth. Overgrazing can cause a severe setback to range and pasture conditions.

Another example of the effect of unbalance is shown in the pruning of apple trees. Careful and limited pruning tends to keep the root and top growth in balance and maintain a condition that encourages fruit production. Severe pruning results in an oversupply of nitrogenous materials from the root system and encourages the development of new shoots with the result that fruit development is discouraged. An excess of carbohydrates over nitrates in the leaves may encourage fruit production.

In transplanting trees, a large portion of the plant's roots are destroyed. A recommended practice observed by nursery men is to prune the top growth to equalize it with the root system of the transplanted tree.

Figure 6-8: Well-pruned Orchard. Careful pruning, to keep roots and top growth in balance, encourages fruit production.

Functions of Stems

The basic framework of the top growth of a plant is provided by its stems. They support leaves, flowers, and fruit. They serve the very important function of transporting water and mineral salts from the root system to the leaves. Young stems which possess chlorophyll cell bodies can also manufacture sugar through photosynthesis. Stems provide the plant with a major storage place for food. Plant food manufactured in leaves and other food-producing areas is distributed through the *phloem*. The phloem is a food-conducting tissue consisting of sieve tubes, companion cells, and fibers. Movement in the phloem is usually downward, from regions of excessive supply to regions of lesser supply. For instance, food may move from the leaves to a place of storage such as roots and stems, or to important growth regions. The phloem system on tree-type plants is just outside the cambium in the bark layer. Any severe injury to the bark layer may destroy the phloem food-transportation system and the plant may die.

Water enters plants mainly through the hairs of the root system. It is conducted upward through the *xylem* of the roots and stems to the leaves. The xylem is the water-carrying tissue of the plant.

Unusual Types of Stem Growth. The potato tuber is an enlargement of the ends of underground stems. Onions or bulbs are modified forms of shoot growth at the base of the plant. *Corms*, the name given to gladiolus and similar plant reproductive stock, are the enlarged bases of stems.

Many plants can be propagated or reproduced from pieces of stems, provided that these pieces include one or more "nodes," stem-joints, with their buds. A particular nuisance to gardeners and farmers is the underground stems of weeds such as thistles. These underground stems are generally rich in stored food. One form of the underground stem is called a *rhizome*. Small pieces of these can start new plants. Quack grass is a plant with rhizomes.

Most gardeners are familiar with the runners of strawberry plants. These are slender stems that grow along the surface of the ground. New shoots or roots develop at the nodes of the runners.

Summary

Only by knowing how plants grow can a farmer intelligently aid their growth.

The life processes of plants are carried on in millions of tiny living cells. The three major parts of a cell are the (1) cell wall, (2) the protoplasm, and (3) the vacuole. The protoplasm contains the *nucleus* and *plastids.* Plastids are classified according to their color as leucoplasts (colorless), chloroplasts (green), and chromoplasts (red or yellow).

Photosynthesis is the process, taking place in the chloroplasts of leaves, by which the energy of sunlight is used to convert water and carbon dioxide into sugar and oxygen.

Transpiration is the process by which plants release water in the form of vapor through pores or stomata in the leaves.

Respiration (oxidation of organic substances) is the process by which plants release the energy stored by photosynthesis. Carbohydrates and fats are broken down, under the regulation of enzymes, into carbon dioxide and water, and energy is made available to the plant.

Digestion is the process by which plants convert food reserves stored after photosynthesis into forms that permit respiration to take place.

Assimilation is the process by which plants convert food into protoplasm and cell walls.

The two major processes associated with *growth* and *reproduction* are (1) the vegetative development of roots, stems, and leaves, and (2) the reproductive development of flowers, fruits, and seeds.

Plant growth takes place as a result of cell division and elongation in the *meristem* regions.

Roots absorb water and mineral salts into the *root hairs* by the process of *osmosis.* They anchor the plant to the soil. They conduct food manufactured by the leaves throughout the root system. They also store food. Root systems are extremely extensive. It is important to maintain a balance between top growth and root systems.

Stems provide the top growth with a framework. They transport water, minerals, and food through the *phloem* and *xylem.*

Young stems manufacture food. Stems store food and often have a reproductive function.

Study Questions

1. Describe and explain the structure and various parts of a plant cell.
2. What takes place in photosynthesis?
3. Explain the processes of transpiration in plants.
4. Explain the processes of respiration in plants.
5. Explain the processes of plant digestion.
6. Explain the processes of assimilation in plants.
7. Explain how and where plant growth takes place.
8. Describe the functions and extent of the plant root system.

Class Activities

1. Observe under a microscope the cell structures in thin slices of plant tissue, such as the potato tuber.
2. Draw diagrams of plant cells you observe under the microscope.
3. Arrange a class demonstration of the process of osmosis. An egg with a tube sealed into the top can be used for this purpose.
4. Observe the stomata of leaves with magnifying instruments.
5. Arrange a class demonstration of the release of oxygen by photosynthesis. Place an inverted glass funnel completely filled with water over a mass of green water plants. Close the top of the funnel. Note how the gas bubbles given off by the plant collect in the funnel tube. Lower a glowing splinter into the funnel tube when filled with gas and note how it bursts into flame. This is a test for oxygen.
6. Measure the amount of water a potted plant needs to keep up its normal growth.
7. Demonstrate how much of the potato tuber is stored-up starch by smearing a cut tuber surface with iodine, which will produce a deep purple color—a test for starch.

Environmental Factors Affecting Plant Growth

Though all major crop plants share the processes of growth described in the preceding chapter, these processes are controlled or affected by the heredity and the environment of specific crops.

Heredity consists of the inborn characteristics of a plant as transmitted by its seed or other reproductive parts, such as stems or tubers. Hereditary factors include the average number of days required for maturity, natural resistance to diseases and insects, potential size and shape, and potential yield.

Environment consists of the external conditions affecting life and development. These include light, temperature, moisture, soil air, such soil conditions as fertility and texture, weeds, diseases, and insects. It is variation in environment that accounts for the fact that when two fields are planted with the same variety of corn, one field might produce 200 bushels to the acre, the other but 20.

This chapter is chiefly concerned with methods for improving the external growing conditions—the environments—of crops.

Plant Environment Applied to Farming

Farmers and soil scientists agree that yields of our common commercial crops can be enormously increased by such improvements in plant environment as fertilization and irrigation. As you may remember from the first chapter, corn yields of 200 bushels per acre are not uncommon on test plots, but the national average yield of corn per acre is about 50 bushels. The farmer who knows what effect environment has on his crops, and what can be done to help his crops grow by improving their environments, will greatly increase crop yields and farm profits.

Effect of Light on Plant Growth. Ordinary farm crops receive their light energy from the sun. Greenhouse plants may have their sunlight energy supplemented by artificial lighting.

Various rays of light influence plant growth in different ways. The rainbow typifies the light rays of infrared, red, orange, yellow, green, blue, violet, and ultraviolet. Plant scientists in control laboratories have experimented with the effects of different light rays on plant growth. For example, plants respond differently to red light than to blue light.

Plant processes and growth affected by light energy are (1) photosynthesis, (2) transpiration, (3) synthesis of proteins, (4) respiration, (5) direction of plant growth, (6) thickness of leaves, (7) time of reproduction or flowering, and (8) form of the shoot system.

Plants can grow in darkness as long as there is a reserve supply of food for them to draw from. Examples are growth of stems on potatoes in storage and of seeds in the ground.

Figure 7-1: Studying the Effects of Light. The top bulb of the "net radiometer" gauges the radiation that comes from the sun; the lower bulb measures the radiation from the corn. The difference is the radiation absorbed by the corn for use during photosynthesis.

Figure 7-2: Results of Light Management. To achieve this remarkable yield, here advertising a hybrid corn, plants were spaced much more closely than usual. Ground leveling, irrigation, manuring, and heavy fertilization more than compensated for diminished light.

Some plants can grow under shaded conditions, whereas others favor bright sunlight. Some plants grow faster at night, provided that night temperatures are not too low. Plants grown under full daylight have sturdier stems and thicker leaves than those grown in the shade. Plants grown under crowded conditions are exposed to less intense light and are taller and more spindly in their growth.

Some plants bloom only when the days are long and the nights are short. They are called "long-day" plants. These include lettuce, clover, radishes, beets, and cereal grains. Some plants normally bloom and set fruit in the autumn when the days are short and the nights are long. They are called "short-day" plants. These include chrysanthemums, asters, dahlias, and cosmos. Some plants appear to be indifferent about the length of the day. Such plants include dandelions, tomatoes, buckwheat, cotton, and sunflowers. For some plants, the length of the light period controls the growth of the shoots which produce the flowering parts.

Light Management. There are a number of practical applications of what we know about the effect of light upon crop growth. One is the practice of using wider spacing for plants. Farmers have achieved remarkable results by planting such crops as wheat and clover in bands, and by wider spacing of such crops as corn. When acreage is limited, other farmers have adopted the practice of very close planting, with the addition of extra amounts of nutrients to make maximum use of the diminished light and to compensate for increased root competition. A further application of our knowledge of the effects of light upon plants is the practice of closely spacing trees in shelter belts or upon former wasteland. Such spacing tends to produce tall, straight, comparatively knot-free trunks, ideal for posts and timber.

Since the discovery of electricity, greenhouse farmers and florists have demonstrated the possibility of remarkably close control over plant size and growth, and time of flower, fruit, or vegetable maturity. By supplementing or reducing natural light, such growers can time bloom or fruiting to take advantage of good market prices.

Effect of Temperature on Plant Growth. Certain temperatures are more favorable to particular plants. Some plants are more tolerant of a wide range of temperature than others. The processes of plant life affected by temperature include: photosynthesis; assimilation; digestion; respiration; absorption of water, mineral salts, and gases; transpiration; secretion of enzymes; permeability of membranes; and rate of movement of materials within the plant. The more active certain of the above processes are, the greater the rate of growth of the plant. Plants vary as to the minimum, optimum, or maximum temperatures which promote growth.

Warm winter daytime temperatures may induce excessive drying out of plant top growth. This loss of moisture by the plant is serious when the plant's root system is absorbing little or no moisture due to frozen soil. Excessive wintertime drying of the top growth may cause winterkilling of trees, shrubs, and berry plants.

Plants absorb less water and soil nutrients at lower soil temperatures, and soil organisms are less active. Sub-freezing temperatures generally will kill young plant growth or destroy in whole or part the fruit or seeds of mature growth.

Cold temperatures slow down or "inhibit" plant growth and physiological processes. Plant scientists have noted that this

Figure 7-3: Temperature Control. Oil-burning heaters like that in the foreground are used with the electric wind machine on the post to protect this California citrus crop from frost. The fan prevents cold air from settling.

slowdown or growth stoppage takes place at a temperature range below 41° and 43° Fahrenheit.

Dormancy is a period of relative inactivity common to many plants and seeds. Dormancy is noted in deciduous trees when they have lost their leaves. Temperate zone deciduous trees must experience a cold spell to break their rest period.

Heaving of soils caused by alternate thawing and freezing may break the root growth of plants. This condition is more prevalent in clay than in gravelly soils. Icing over of fields during the winter months can cause winter loss in perennial plants such as alfalfa.

Spring freezing of buds and flowers is often brought about by a too-early spring emergence of the fruit-bearing parts of the plant. Extreme cold spells with sub-zero temperatures can cause plants to die. This is commonly observed in trees, shrubs, and berry plants which fail to come through the winter.

Extremely high temperatures can slow plant growth and cause a wilted and dried-up appearance. In the seedling stage, a plant may favor a cooler temperature than near the ripening or mature stage. High temperatures greatly increase transpiration.

Temperature Management. What can be accomplished by management of temperatures is again best seen in greenhouses. Greenhouse growers of fruits and flowers are able to speed or delay growth by as much as several weeks by raising or lowering temperature, respectively, and are also able to prolong the period of full bloom or full fruit maturity by as much as a week or more by keeping temperatures cool to take advantage of market conditions.

Temperature control in the field is often achieved by planting crops at different times of the year. Thus small grains are usually planted early in the spring whereas corn or sorghum is planted later.

Another conspicuous example of the application of temperature control to field crops is the use of "smudge pots" by citrus growers to prevent frost. Snow fences, mulches, shelter belts, and fall plowing, which serve to cut the chilling effects of wind and to trap snow, with its ground-insulating qualities, are further examples of temperature control.

Effect of Moisture on Plant Growth. The rate of growth of plants is generally in direct proportion to the available supply of soil moisture. However, too much water may be harmful to

Figure 7-4: Putting Wind to Work. The snow trapped by this "snow fence" of sorghum stubble will not only provide moisture for next spring's growth but is helping protect fall-planted wheat from extreme winter temperatures.

Figure 7-5: Measuring Wind Velocity and Temperature. This USDA scientist studies the effects of wind and temperature on the corn in the background with gauges measuring wind velocities at slightly different heights and with the bottle-like devices for measuring temperatures.

plant growth. The soil moisture supply depends on (1) the amount of rainfall and irrigation and its distribution during the growing season, (2) evaporation, (3) the height of the water table, (4) the rate of water percolation downward in the soil, (5) the soil structure, which favors or hinders the rise of water, by capillarity, from the water table, (6) soil texture, and (7) vegetative covering. Drought is a common cause of poor yields, especially in areas of low rainfall. Extended droughts also occur in areas with fairly heavy rainfall.

Periods of high and low moisture supply may cause cracking in fruits and tubers. This is a condition brought about by alternate periods of rapid and restricted growth.

Prolonged periods of rain in the spring will delay planting operations. Seeds may rot in the ground. Working wet soils may cause them to clod. Prolonged fall rains may delay harvesting, and cause weathering of fruits and seeds which may lower their market value.

Heavy snow cover serves to protect dormant plant growth during the winter months. Snow cover aids moisture retention.

Effect of Humidity, Precipitation, and Wind on Plant Growth. All farmers are aware of the importance of soil moisture on plant growth. Factors which affect favorable moisture conditions for the plant are wind, and humidity, and temperature. Wind increases the transpiration or loss of water from the plant. Humid conditions reduce the vapor or moisture loss from plant leaves. The drier the air, the greater the water loss. Crops take on a wilted appearance when the tops lose water faster than the roots can absorb it. High transpiration loss in plants is due mostly to high temperatures.

Lodging in grain is caused by high winds and heavy rain. Hail damage to grain and other crops is caused by a combination of wind and cooling temperatures at high altitudes. Heavy rains accompanied by high winds can cause physical damage to growing plants. Floods cause physical damage to plants and may smother them.

Moisture Management. All crops are benefited by water management. You have already learned some of the ways by which natural water may be encouraged to enter the soil and remain available for plant use. In later chapters you also will learn how insufficiently watered fields may be irrigated and how over-watered fields may be drained. Moisture management practices include plowing, using green and animal manures, mulching, irrigation, drainage, planting shelter belts, and terracing.

At present, little can be done to compensate for low humidity in the field except to increase amounts of soil water, either by

Figure 7-6: Grass in Rows. Wide spacing of rows permits grass to grow in a dry region. Note cultivation to prevent weeds from competing for the limited moisture.

conservation practices or by irrigation, though protection against wind may raise humidity to an extent. Wind as a drying agent can be controlled by shelter belts, and to some extent by mulching. Reduction of wind speed also is of great importance in cutting the loss of water and soil during storms, since it lessens the force and the angle of raindrops, as will be discussed in more detail in the chapters on conservation.

Effect of Soil Air on Plant Growth. Roots require air for respiration. Poor soil aeration results in decreased root absorption of water and nutrients. Most plants except those adapted to wet conditions cannot live long in water-soaked soil. Soils tightly packed or with an excess of water have insufficient soil air for plant growth. Plants grown in well-aerated soils are generally taller and heavier, and have a more extensive root growth.

Soil Air Management. The amount of soil air may be increased temporarily by tillage, permanently by drainage (if necessary) and by the use of green and animal manures.

Effect of Soil Texture and Structure, Fertility, and Reaction on Plant Growth. Almost all commercial crops prefer soil of good texture and structure, high fertility, and neutral to slightly acid reaction. Methods of improving these are discussed in detail in the chapters that follow.

Effect of Weeds on Plant Growth. Weeds compete with most of our commercial crops for light, water, and nutrients. To help

Figure 7-7: Rice Terraces. Rice is one of the few commercial crops that can grow in standing water. Most crops require good soil drainage to provide aeration.

control weeds, seed that is free from impurities should be used. Growing weeds may be controlled by tillage or by application of various new chemical weed-killers. Since the latter are sometimes harmful to crops if improperly applied, farmers are best advised to take advantage of the knowledge and, if necessary, of the actual assistance of persons familiar with their use. It is also important to check the possibility that the use of certain chemicals may be dangerous or illegal for some or all purposes, because harmful to farm animals or to human beings.

Effect of Insects and Diseases on Plant Growth. It is best to take advantage of hereditary immunity or resistance to insects and diseases when this is possible. There are, of course, very many commercial preparations which will control plant diseases and insects; the farmer may choose those adapted to local conditions such as rainfall, and those most convenient to apply.

Figure 7-8: Contour Strip Cropping. Good soil management increases the resistance of crops to damage by insects and diseases.

Figure 7-9: Plant Growth Laboratory. This controlled environment laboratory enables scientists to experiment with light, temperature, and other environment factors that affect plant growth.

Farmers should remember, however, that very many plant ailments are caused by soil deficiencies of fertility or moisture, and that others are encouraged by such deficiencies. Poor soil management practices lower the natural resistance of plants to infection. Good fields tend to produce healthy crops.

Summary

The processes of plant growth are controlled or affected by heredity and environment.

Heredity consists of the inborn characteristics of the plant.

Environment consists of the external conditions affecting life and growth. Yields of our common commercial crops can be enormously increased by improvements in plant environment.

Many essential plant processes are affected by the environmental condition of *light*. Ordinary farm crops receive their light energy from the sun. Farmers apply our knowledge of the effects of light upon growth by such practices as wider spacing of crops, planting in bands, and close-planting with extra fertilization.

Processes of plant life are also affected by *temperature*. Early and late planting, snow fences, shelter belts, and mulches are examples of temperature control.

The rate of growth of plants is generally in direct proportion to the available supply of soil *moisture*. Humidity, precipitation, and wind are other factors, connected with moisture, that affect plant life processes. Moisture management includes plowing, using green and animal manures, mulching, irrigating, draining, planting shelter belts, and terracing.

Soil air is another environmental factor affecting growth. Soil air may be increased by plowing, using green and animal manures, and draining overwatered fields.

Other environmental factors affecting growth are *soil texture, structure, fertility*, and *reaction; weeds;* and *insects* and *diseases*. There are various ways of managing these conditions.

Study Questions

1. What is the effect of temperature on plant growth?
2. What is the effect of light on plant growth?
3. What is the effect of soil temperature on plant growth?
4. What is the effect of soil air on plant growth?
5. What is the effect of humidity, precipitation, and wind on plant growth?
6. What is the effect of soil moisture on plant growth?

Class Activities

1. Take a field trip to lawns and gardens to observe plant growth affected by light, temperature, and the season of the year.
2. Bring from home potted plants showing effects of environmental factors.
3. Determine with potted plants the effects of too much, too little, and optimum amounts of soil water.

Elements Essential to Plant Growth

An important part of the study of the life processes of plants is a knowledge of the *elements* (basic chemical substances) essential for plant growth. Plant scientists recognize 16 elements as essential to all or some plants. Important elements from air and water include (1) carbon, (2) hydrogen, and (3) oxygen. Thirteen essential items are obtained by plants from soil. Essential soil elements used by plants in relatively large quantities include (1) nitrogen, (2) phosphorus, (3) potassium, (4) sulfur, (5) calcium, and (6) magnesium. Essential soil nutrient elements used in extremely small quantities include (1) iron, (2) boron, (3) zinc, (4) manganese, (5) copper, (6) molybdenum, and (7) chlorine.

Farmers and gardeners may easily mistake hunger signs in plants for lack of moisture. Plant starvation may be the cause when leaves turn yellow and die or drop prematurely.

Air and Water As Plant Nutrients

Some 300 years ago, Jan Van Helmont, a Belgian scientist, planted a five-pound willow tree in a box of 200 pounds of soil. Having carefully watered the plant during five years, he discovered that it now weighed 169 pounds, while the soil had lost only two ounces in weight. At the time, Van Helmont decided that plants need only water for growth. Scientists have since discovered that, in addition, they require the carbon dioxide in air to manufacture food by photosynthesis. Scientists suggest that the two-ounce loss in original soil weight supplied the small amount of other elements necessary for plant growth.

Plants obtain carbon from the air in the form of carbon dioxide (CO_2), hydrogen from water (H_2O), and oxygen from both water and air. Air, containing about 78 per cent nitrogen and about 21 per cent oxygen, has 0.03 per cent carbon dioxide.

The Importance of Carbon Dioxide. Up to 95 per cent of the dry matter of plants is made up of carbon compounds, which include starches, cellulose, fats, oils, and lignin. As you have learned, this carbon is supplied by carbon dioxide, a gas in air. Used by green plants in the process of photosynthesis, carbon dioxide dissolves in the cell-solution water to form carbonic acid.

Plant scientists have secured increased yields of crops when the percentage of carbon dioxide in the air has been increased. They have demonstrated that part of the value of fertilizers and manures comes from the fact that these increase the activity of soil microorganisms, which release carbon dioxide. By this means, carbon dioxide is made available in larger quantities above the soil surface. Some scientists estimate that the total amount of carbon dioxide released into the atmosphere by the action of soil

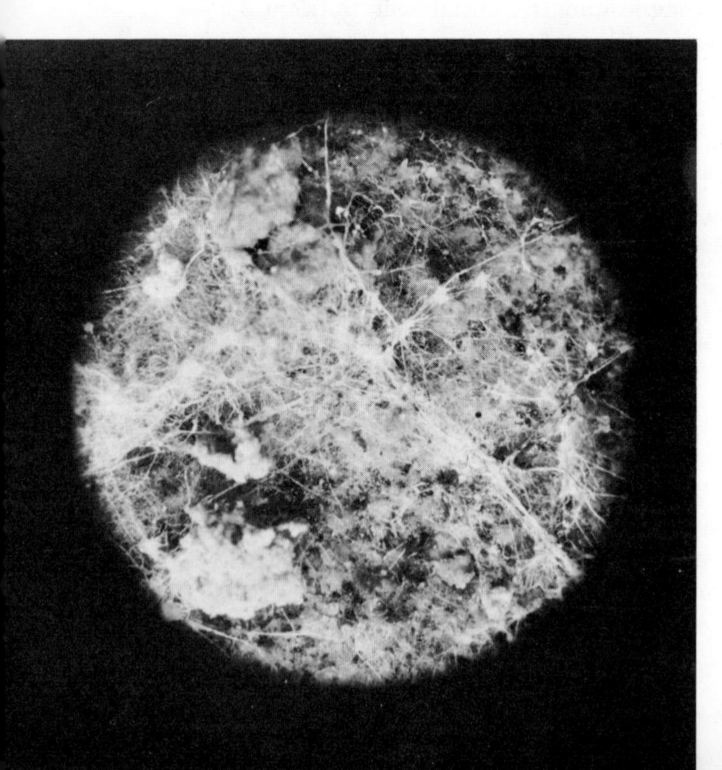

Figure 8-1: Soil Fungi. Soil microorganisms such as these, here magnified 12 times, release carbon dioxide into the soil air.

bacteria exceeds the amount released by all animals. This estimate is based on the fact that the percentage of carbon dioxide in the soil air is greater than in the atmosphere. Carbon dioxide is a by-product of the mineralization of organic matter by soil microorganisms.

Oxygen. The element oxygen comprises about 21 per cent of the atmosphere. Oxygen performs an essential function in the *respiration* of plants. Oxygen and hydrogen make up water (H_2O), which is an essential compound, along with carbon dioxide, for the manufacture of carbohydrates through photosynthesis. Both carbon dioxide and water furnish oxygen to plants.

Hydrogen. A principal constituent of plant materials, along with carbon and oxygen, is *hydrogen*. Water is its source as a plant nutrient element.

Essential Major Soil Elements

Nitrogen. With enough water and suitable temperatures, the rate of growth of plants is largely attributed to the available nitrogen in the soil. An early indication of nitrogen deficiency is a

Figure 8-2: Texas-sized Corn. Once plants have enough water and suitable temperatures, their rate of growth is largely determined by the amount of nitrogen in the soil.

yellowing of leaves and a stunting of growth, though yellowing of leaves sometimes may be due to other causes. An excess of available nitrogen is commonly associated with too-vigorous vegetative growth, and lodging (susceptibility to being beaten down, as by rain) in grain. It also will lessen food storage in the plant and hinder fruit and seed development.

Nitrogen starvation in corn is indicated when the lower leaves turn yellow, with most intense yellowing along the midribs. Farmers may mistakenly blame this condition on drought, because nitrogen starvation may be more severe during dry periods. Unfilled corn ears or those with just a few kernels near the butt are further indications of nitrogen starvation.

In the case of small grains and similar field crops, nitrogen starvation is commonly indicated by a light green to yellow color of the whole field.

Nitrogen deficiency is apt to show up in crop rotations when too much straw or corn stalks have been plowed under. This is thought to be caused by a temporary free-nitrogen shortage, in which soil microorganisms have tied up the available soil nitrogen supply while breaking down the fresh organic material.

Nitrogen is a strategic item in the formation of protoplasm, and is an essential element in chlorophyll. Plants do not use the "elemental" nitrogen which constitutes about 78 per cent of the atmosphere; their chief source of this element is the breakdown of organic matter. Plants can also utilize nitrogen in the form of ammonium salts. Barnyard manure is an important source of nitrates and ammonia.

Nitrogen is an essential constituent of proteins. It is taken up by the plant in the form of ammonium (NH_4) or as a nitrate such as calcium nitrate, $Ca(NO_3)_2$. In the living parts of the plant, the nitrogen compounds are combined with carbohydrates, such as sugars and starches, to form amino acids. These amino acids are the basic constituents of proteins.

The main commercial nitrogen fertilizers are (1) nitrates, (2) ammonia compounds, (3) organic compounds such as tankage and cottonseed meal, and (4) those of the amide form such as urea and calcium cyanamide. The organic nitrogen compounds must be converted into available forms by the action of soil microorganisms.

Phosphorus. Over 50 per cent of the phosphorus in a mature plant may be located in the fruits and the seeds; generally it is most abundant in the young growing parts of the plant. Phosphorus is a component of some plant proteins. The phosphorus requirements of annual plants are most heavy during the first few weeks after germination and again during the fruit and seed formation period. Phosphate fertilizers encourage root growth and thus aid root crops such as beets. Phosphate also helps hasten the maturity of grain crops.

Phosphorus deficiency shows up as stunted top growth and root growth, spindly stalks, and delayed maturity. The leaves of some phosphorus-deficient plants have a purplish color, and tips of older leaves often die. Poor fruit and seed development also are often caused by lack of phosphorus.

Phosphorus starvation in corn is most common when plants are just a few inches tall. The lower leaf blades turn purple first at the tips. During cold, wet weather, the young plants may become almost entirely purple. The purplish color may disappear when the roots become better established, but thereafter the lower leaves of the corn plant may turn yellow during the time of ear formation.

Phosphorus starvation in alfalfa shows up as a yellowing and dropping of lower leaves just before and at the time of blossoming. Fertilizer corrects the condition.

Potassium. Buds, young leaves, root tips, and other growing parts of the plant are rich in potassium, while older tissues contain relatively little. Older leaves and other parts of the plant lose potassium which is transported to the growing regions of the plant. Unlike nitrogen and phosphorus, potassium does not seem to be used as a part of the organic materials of the plant, such as protein. Its exact functions are somewhat obscure, but appear to be chiefly regulatory. Potassium is the most abundant element in the mineral ash of the plant. Plants may contain in their mineral matter about ten times the percentage of potassium in soil. About half the mineral matter of plants is potassium.

Plants do not grow normally in soils deficient in potassium. Many plant diseases and low yields are traced to lack of it. Potassium fertilizers are potassium chloride, potassium sulfate, potassium phosphate, and potassium nitrate.

Potassium deficiency is noted in stunted growth. Leaves may appear to be dull green. Their edges and tips are often scorched, and bronze-colored spots develop. Older leaves are affected first.

Potassium starvation in alfalfa shows up in the formation of a pattern of white dots on the leaves. These white dots are sunken spots without any color. The yellow leaflet edges, characteristic of potassium starvation in many plants, may not always be present. Yellow edges on leaves may also be caused by insect injury.

Mature corn ears that are filled out but have chaffy and soft-tipped kernels indicate potassium starvation.

Sulfur. Sulfur is found in most plant tissue. It is an important constituent of proteins. Much of the sulfur in plants remains in inorganic forms such as sulfates. Sulfur in the soil appears to favor root growth of some plants. Chlorophyll development is favored by sufficient sulfur in the soil. The pale green color of leaves changes to a deep green when sulfur is supplied to deficient soils. A lack of sulfur is also thought to retard the development of nitrogen-fixing nodules on legumes. Sulfur may be added to soils with fertilizers such as ammonium sulfate.

Figure 8-3: Potassium Deficiency. A deficiency of potassium accounts for the poor alfalfa growth in the center foreground.

Figure 8-4: Liming Pasture. Lime materials are compounds of calcium, which is essential to plant growth.

Calcium. The value of calcium compounds or *lime materials* in neutralizing soil acidity is well known. Calcium itself is an important plant constituent. Lack of it results in a stunting of root growth in many plants. The major portion of calcium in most plants is in the leaves. Some tests indicate that there is about eight times as much calcium in the mineral matter of plants as in soil. Calcium performs various functions in the life processes of the plant.

Signs of calcium hunger in plants include a tendency to shed blossoms and buds prematurely. The younger parts of the plant die prematurely. The leaves become hard and stiff and often yellow. Mottling or brown spots on the leaves are common. The margins of younger leaves have a scalloped appearance.

Magnesium. A lack of magnesium results in pale, sickly foliage, an unhealthy condition called *chlorosis*. The chlorosis or sickly appearance of the leaves affects first the older leaves and later the younger ones. The top of the plant may be green and the lower portions yellow. The veins of the leaf may appear green and the leaf web tissue yellow or white. Magnesium is the only mineral element contained in the chlorophyll of plants.

Figure 8-5: Irrigating Beets. Irrigation water sometimes contains boron, which is essential to plant growth in very small quantities but harmful if highly concentrated.

Essential Minor or Trace Soil Elements

Nutrients required by plants in large quantities are nitrogen, sulfur, phosphorus, potassium, magnesium, and calcium. Scientists have established that other elements in minute quantities are essential to plant health and growth. For a long time, these elements were not identified as essential because the quantities needed were so small. Yet whole fields and orchards of agricultural crops may suffer from the lack of these minor elements. The essential minor elements are iron, boron, zinc, manganese, copper, molybdenum, and chlorine.

Iron. Plants contain larger amounts of iron than of the other trace elements. Soils normally have an abundance of iron for plant needs, but it is more available for plant use in acid soils than in neutral and alkaline soils. Iron is an important element in aiding the process of photosynthesis. Plants lacking in available iron are affected by chlorosis of leaves. Younger leaves are first affected. The veins of the leaf may remain green and the leaf web tissue may turn yellow or white. Affected leaves are curved in an upward direction.

Boron. Boron deficiency results in darkening of tissues and various growth abnormalities and disturbances, including (1) internal cork of apples, (2) top rot of tobacco, (3) cracked stem of celery, (4) browning of cauliflower, (5) heart rot of sugar beets, and (6) "monkey face" of olives. Application of ten to 20 pounds of borax per acre is usually sufficient to cure these.

Excess quantities of boron in the soil are injurious to plants. Too much boron may be added to soils through irrigation water. Boron compounds in high concentrations are used as weed-killers.

Zinc. Certain disease conditions have been corrected by applications of zinc. "Little leaf," a disease of deciduous trees (those that lose their leaves in the fall), and "mottle-leaf" of citrus trees are cured by spraying the trees with zinc salts, by injecting weak solutions of zinc salts into the trunks, or by driving zinc brads into the trunks. Soil treatment can solve zinc deficiency.

Manganese. Lack of manganese causes chlorosis of the younger leaves at the top of the plant. The veins of the leaf may remain green and the tissue between becomes light green, yellow, or almost white. The lower leaves remain green. By spraying or dusting, crops showing manganese deficiency can be cured with as little as 20 pounds of manganese sulfate per acre. Soils having excess amounts of manganese are injurious to plants.

Copper. Beneficial effects have been observed in plants sprayed with copper compounds. Applications of copper have corrected abnormalities in plants grown on peat, for instance. Plant foliage showing a deficiency of this element has a bleached appearance. "Dieback" of new growth in citrus-growing areas is associated with copper deficiency, as is the presence on mature citrus fruit of reddish-brown abnormal growths. An excess of copper in soils is equally injurious.

Molybdenum. Molybdenum is considered essential to some plants. For example, it has been found to be required for the normal growth of tomato seedlings. Molybdenum-deficient plants are stunted and yellow in color. Molybdenum is needed by organisms on the roots of plants for nitrogen fixation. Plants containing an excess of molybdenum are injurious to animals.

Chlorine. Chlorine is the most recently recognized essential element. It is picked up by winds from sea spray and enters the soil in rainwater. Under field conditions, chlorine seldom is a limiting growth factor. During laboratory tests, chlorine deficiency is shown by wilting, chlorosis, and necrosis of leaves.

Nonessential Soil Elements

Plants possess a high degree of power to select the food elements that they need. Elements which are in plentiful supply in a soil may be taken up in larger amounts. Some plants are more selective than others.

Two elements often found in plants but not essential to them are iodine and cobalt. Both of these are essential to animal life, however, and if the plants animals eat are lacking in them, the animals will display deficiency symptoms. For example, goiter in human beings is caused by insufficient iodine in the diet. Scientists have identified as many as 60 other elements in plants, including fairly large quantities of silicon and sodium.

Figure 8-6: Cobalt Deficiency. The stunted lamb at right was raised on hay from land deficient in cobalt.

Figure 8-7: Applied Hydroponics. Hydroponics is used extensively in research on plant nutrition. The equipment shown is used for growing plants in growth chambers for livestock feed, but the value of the practice has not been generally accepted.

Summary

Plant scientists recognize 16 elements as essential to plant growth. Elements are basic chemical substances. Farmers easily mistake hunger signs in plants for lack of moisture.

Important elements obtained by plants from air and water include (1) carbon, (2) hydrogen, and (3) oxygen.

Essential soil elements used by plants in relatively large quantities are (1) nitrogen, (2) phosphorus, (3) potassium, (4) sulfur, (5) calcium, and (6) magnesium. Essential minor or *trace* elements used by plants include (1) iron, (2) boron, (3) zinc, (4) manganese, (5) copper, (6) molybdenum, and (7) chlorine.

Deficiencies in any of these soil elements produce characteristic symptoms in various crops. Plants also become subject to infection by certain diseases when soils are deficient in different elements.

Plants also absorb nonessential elements. While not needed for plant growth, these are often essential to human beings and animals.

Study Questions

1. What nutrients does a plant secure from the air?
2. What is the importance of carbon dioxide for plant growth?
3. What are the functions of oxygen and hydrogen for plant growth?
4. For each of the essential elements the plant absorbs from the soil, indicate (1) its effect upon plant growth, and (2) hunger signs.
5. What elements are found in plants which are considered nonessential for plant growth, and of what value are they to humans and animals?

Class Activities

1. Observe the root growth on pieces of vine in a glass of water. Note that ordinary tap water will contain some minerals in solution which plants ordinarily obtain from the soil. How might this fact have affected the result of Van Helmont's experiment? Consider, also, the effect which dust might have had on this experiment. Discuss ways in which the experiment might be reperformed with greater accuracy.
2. Arrange to have pots of growing grain watered with nutrient solutions of different amounts and combinations of nitrogen, phosphorus, and potash fertilizer, and observe differences in growth. Plants grown in sand will respond quickly to nutrient solutions.
3. Summarize the various "hunger signs" in crops.

Soil Reaction—Acidity and Alkalinity

Students of soils need to understand soil acidity and alkalinity to appreciate the problems involved in plant nutrition. Knowing that a soil is acid or alkaline is helpful in choosing management practices, such as adding lime. Soil scientists speak of soil acidity and alkalinity as *soil reaction.*

To understand this subject, you first must learn the chemical and physical processes that are involved, and the symbols with which these processes are described.

Understanding Ionization of Molecules. To understand the chemical nature of acidity and alkalinity it is necessary to understand ionization of molecules of matter. The smallest amount of matter that can exist by itself and retain all the properties of the original substance is a *molecule.* A row of 80 million water molecules is about one inch long.

Molecules of a substance contain the same number of positive and negative *ions,* which are electrically charged parts. When dissolved in water, molecules of many substances break into two or more of these ions. Ions derive their charge from either an excess or a deficiency of electrons. An *electron* is a unit of negative electricity. An excess of electrons gives an ion a negative charge, whereas a loss of one or more electrons leaves the ion with a positive charge.

The acidity or alkalinity of the soil solution is determined by the relative number of the hydrogen (H^+) ions and hydroxyl (OH^-) ions.

Most soil solutions are highly diluted and water is their principal constituent. They also contain dissolved plant nutrients, mostly in the form of ions. For example, most of the potassium chloride in a soil solution exists as potassium ions and chloride ions. The potassium ion has lost an electron and has a positive (+) charge and is called a *cation.* The chloride ion has gained an electron and has a negative (−) charge. It is called an *anion.*

Understanding Acidity and Alkalinity. In a solution, there is a balance between the number of hydrogen (H^+) ions and hydroxyl (OH^-) ions. It is impossible to decrease the number

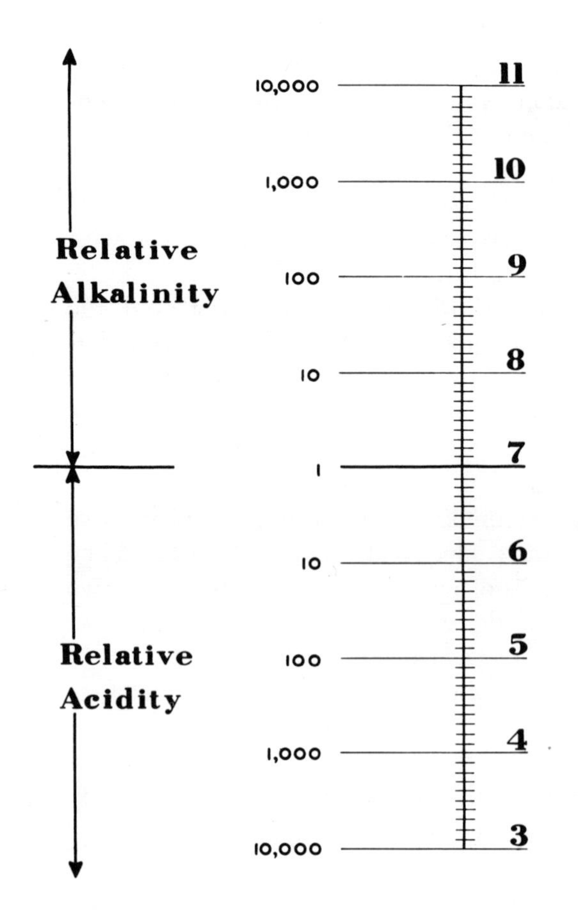

Figure 9-1: Relative Acidity and Alkalinity. You can use this chart to determine degrees of soil acidity and alkalinity. For instance, **pH** 6 is shown to be 10 times as acid as **pH** 7, neutral. How many times more alkaline than **pH** 7 is **pH** 10?

of one of these ions without increasing the number of the other ions, just as when one end of a seesaw goes up, the other must go down. When a solution contains more H^+ ions than OH^- ions, it is acid. When the OH^- ions are more abundant, the solution is alkaline. A neutral solution has an equal number of H^+ and OH^- ions.

When the water molecule itself ionizes, it forms one hydrogen ion and one hydroxyl ion, according to the chemical formula: $H_2O \rightarrow (H^+)(OH^-)$.

From this reaction, some might expect a soil solution to have the same number of hydrogen and hydroxyl ions and be neutral in reaction. However, either the soil solids or salts dissolved in the water of the solution provide additional H^+ or OH^- ions, so that the H^+ and OH^- ions are seldom present in equal numbers.

Expressing Soil Acidity and Alkalinity

Soil scientists speak of soil acidity and alkalinity, you may remember, as *soil reaction.* Soil reaction is expressed in *p*H values. The "*p*" of the symbol refers to "pressure" and the "H" to "hydrogen ions." Thus, "*p*H" refers to the *pressure* or *concentration of hydrogen ions.*

The *p*H scale represents the degree of acidity or alkalinity of solutions.

Freshly distilled water is neutral and typifies the balance of hydrogen (H^+) ions and hydroxyl (OH^-) ions. An alkaline soil is one which contains an excess of OH^- ions over H^+ ions in the soil solution, an acid soil one with an excess of H^+ ions over OH^- ions. In soils with a balance of H^+ and OH^- ions, the *p*H is 7, neutral.

Soils which are weakly acid have the acidity of fresh milk, or *p*H 6.6 to 6.69. Soils strongly acid have acidity of sour milk, or 4.6 to 4.8. Extremely acid soils have the acidity of lemon juice, *p*H 2.2 to 2.6. Soils are seldom this acid.

Soils that are weakly alkaline have the alkalinity of human blood, 7.35. Soils that are weakly alkaline to alkaline have the alkalinity of sea water, 7.5 to 8.4. Soils that are strongly alkaline have the alkalinity of a soap solution, 8.0 to 10.0.

Figure 9-2: The Colloid Complex. Rich loam such as that illustrated has a high proportion of very fine particles of clay, organic matter, and oxides. Can such a soil store many nutrients or few?

Soil Particles and Reaction

The important chemical properties of acidity and alkalinity, or reaction of soils, are bound up in the minute clay particles called *colloids.* Colloid materials are so fine that they can remain in suspension in solutions with water or other fluids, rather than settling at the bottom. These clay particles may consist of clay minerals, fine particles of organic matter, and oxides of iron, aluminum, or silicon.

The minute clay colloid materials are composed largely of minerals developed in the soil. Thus the original minerals that made up the soil parent material have gone into solution and emerged as new minerals or compounds which are characteristic of the particular soil. *Colloid particles* have properties which hold the soil particles together. Fine clay colloid particles also perform an important function in holding on to certain plant nutrients. Plant nutrients are stored in the *colloid complex.* These stored nutrients are made available for plant use under different conditions.

Fine clay mineral particles consist largely of silicon, aluminum, and water. Other elements include iron, calcium, magnesium, potassium, and sodium. The particles are similar in form to sheets of mica.

The individual sheets which make up the mica-like clay particles sometimes are far enough apart so that nutrient and hydrogen ions can be held between sheets, more or less like food on the shelves of a grocery store. In other clay minerals the sheets or shelves are too close together to allow storage between sheets, and all available nutrients must then be attached to the outside surface.

The fine clay particles are the "home plates" of hydrogen or hydroxyl ions as well as of soil nutrient ions such as calcium, potassium, and phosphorus.

Figure 9-3: Sandy Soil. Loose, poorly aggregated, sandy soil tends to have a low reserve acidity. What kind of soil tends to have a high reserve acidity, according to page 128?

Clay particles are usually negatively charged and consequently behave like negatively charged ions, although they are much larger than ions. They attract positively charged ions and tend to hold them adjacent to their surface. The most common positively charged ions surrounding the particles are hydrogen, calcium, magnesium, potassium, and sodium.

When hydrogen (H^+) ions are abundant around clay particles, they tend to *ionize* in (be liberated into) the soil solution and consequently make the solution acid. When any of the other ions ionize from the clay surface, the clay tends to attract hydrogen ions from the solution to neutralize its negative charge. This reduces the hydrogen ions in the solution and consequently tends to make the soil solution low in hydrogen ions, or alkaline. Thus when hydrogen ions are dominant, the soil will be acid. When other ions are abundant on clay particles, the soil will be alkaline.

Active and Reserve Acidity. Tests for acidity express the concentration of free hydrogen ions in the soil solution. This concentration of *free* ions is called *active acidity*. The soil still contains a *reserve* supply of hydrogen ions which are held by minute soil colloids.

A *p*H value expresses the active acidity of the soil, then, but does not indicate the *reserve acidity* held in *colloidal suspension*. In other words, *p*H is not a measure of total acidity. Soils with a high clay content require more lime to take care of the potential acidity of the soil than soils which contain less clay. Soil scientists for this reason have recommended different quantities of lime to displace the excess hydrogen ions for soils differing in texture.

Buffering in Soils. When hydrogen or hydroxyl ions in the soil solution are neutralized, the ions held on the colloidal particles tend to ionize off and take the place of the ions neutralized in the solution. Thus the ions held on soil particles make the soil solution resist rapid changes in *p*H. This ability of the soil solution to resist rapid changes in reaction is known as *buffering*. The more ions held on the colloidal particles, the more highly the soil is buffered. Thus soils high in clay or organic matter are highly buffered. Sandy soils are only slightly buffered.

A highly buffered acid soil requires a heavy application of lime to reduce its acidity or make it alkaline. Similarly, a highly buffered alkaline soil is difficult to make acid.

Figure 9-4: A Region of High Rainfall. This well-managed West Virginia field tends to be acid because a high annual rate of precipitation leaches out "basic" ions such as calcium ions, and leaves hydrogen ions in their place. Note that these fields, if left undisturbed, would soon revert to forest.

Formation of Acid and Alkaline Soils

Soils in areas of high rainfall are usually acid. In dry areas, soils are usually alkaline. Soils of the eastern half of the United States are predominantly acid. Most of the soils in the West are alkaline.

In humid areas, soil moisture moves through the soil and out as drainage water. Hydrogen ions of water gradually tend to replace other positive ions, such as those of calcium, attached to the clay particles. Thus the "basic" ions such as calcium tend to be leached out and hydrogen ions take their place. In dry areas, almost all water that enters the soil is again lost from the surface by either direct evaporation or transpiration through plants. Thus in dry areas there is almost no leaching of basic ions and the soil tends to remain alkaline. Usually 20 to 30 inches of rainfall are enough to cause the formation of acid soils.

Undisturbed or virgin soils in humid areas are usually covered with forest. In dry areas, grass is the predominant natural cover. Because of the prevailing high rainfall, naturally forested soils are usually acid. Soils which had a grass cover prior to cultivation are usually alkaline.

When plant growth is removed from the land, as in harvesting, the basic ions which the plants removed from the soil are lost from the soil rather than returned through processes of decay. This also tends to make the soil acid. Legume plants are heavy users of basic ions and consequently tend to increase soil acidity. Some fertilizers leave acid residues. Organic acids are also produced in the decay of organic matter, and these tend to increase acidity.

Range of Reaction in Soils. In regions of the United States where the average rainfall is 25 inches or more, soils will rarely exceed *p*H 8. The usual *p*H range of soils in humid areas is from *p*H 5 to *p*H 6.8. This is described as an "acid to weakly acid range." Rarely do soils have a *p*H 4, which is considered strongly acid.

Western soils are generally alkaline. In dryland areas of our Western states the average rainfall range is from 10 to 20 inches. In these areas there is considerable surface evaporation of water. These soils are usually less alkaline than *p*H 8.6, unless they have been adversely affected by sodium salts, when the *p*H may go to 9.5 or even more.

Seasonal Variations in Acidity. Soil scientists have observed that soils tend to increase slightly in acidity during the summer. During the late fall and winter, soils gradually return to the *p*H level of early spring. As plants grow rapidly during the summer they absorb the soluble salt nutrients and permit the soil to release more acid or hydrogen ions. As organic matter is decomposed during the summer, it releases nitrates and sulphates which form small amounts of nitric and sulfuric acid. The seasonal variation in acidity is not considered significant in most soils. Sandy soils have a low clay content and may be affected more by the seasonal variation in acidity. Such soils would respond quickly to a light lime application.

Figure 9-5: Reaction and Growth. Each tube, labeled according to the reaction of the solution it contains, is filled with identical nutrients dissolved in water. Can you tell which two reactions this crop seems to favor?

Reaction and Plant Growth

Some plants require strongly acid soils. Many plants can survive under highly acid soil conditions but would grow better under more favorable conditions. Other plants grow well on acid soils which are well supplied with plant nutrients and moisture. Plants which are found growing on acid soils are usually more tolerant of soil acidity.

Generally, crops favor a slightly acid soil. Soil acidity that crops favor is also favorable to the activities of soil microorganisms. These soil microorganisms help make plant nutrients available for plant use. When excess acidity is corrected by the addition of lime to soils, the calcium contained in lime is used abundantly by plants.

Some soils which formerly grew abundant clovers have later been taken over by other types of plants as a result of a change in the soil condition. These new plants are called *invaders*. Some plants, such as sorrel, which are frequently associated with acid soils, grow better in soils which have been limed. Experimental studies have shown that most crops will grow better in a soil that is slightly acid than in a neutral or alkaline soil. Even an acid-sensitive crop like alfalfa does better on a slightly acid soil. This would suggest that *an excess of lime applied to a soil is harmful.*

Plants that thrive best on strongly acid soils include the cranberry, blueberry, rhododendron, watermelon, and azalea. Acidity in soils can be increased by the addition of (1) ground sulfur, (2) iron sulfate, or (3) aluminum sulfate. The use of sulfur is the most common practice for increasing soil acidity. The mixing of acid peat with soil is another way to increase acidity.

Farm crops may grow on soils with varying degrees of acidity. Plants which grow on acid soils may do better on less acid soils. The following listing indicates farm crops which tolerate varying degrees of acidity.

Degree of Acidity Which Legumes Favor or Tolerate

Weakly Acid Soil	*Moderately Acid Soil*	*Acid Soil*
Alfalfa	Red clover	Vetch
Sweet clover	Alsike	Crimson clover
Ladino clover	White clover	Field beans
	Peas	Kudzu
	Lima, pole, and snap beans	Lupine
	Soybeans	Velvet beans

Degree of Acidity Which Common Field Crops Favor or Tolerate

Weakly Acid Soil	*Moderately Acid Soil*	*Acid Soil*	
Sugar beets	Cotton	Barley	Buckwheat
	Peanuts	Corn	Sorghum
	Wheat	Oats	Millet
		Rye	Redtop
		Tobacco	

Degree of Acidity Which Vegetables and Similar Crops Favor or Tolerate

Weakly Acid Soil	*Moderately Acid Soil*	*Acid Soil*
Cabbages	Carrots	Potatoes
Cauliflower	Cucumbers	Sweet potatoes
Lettuce	Brussels sprouts	Parsley
Onions	Kale	Strawberries
Spinach	Kohlrabies	
Asparagus	Pumpkin, squashes	
Beets	Sweet corn	
Parsnips	Tomatoes	
Rutabagas	Turnips	
Celery	Radishes	
Muskmelons		

Crops Which Favor Strongly Acid Soils

Blueberries	Cranberries	Watermelons

Calcium Supply in Acid Soils. In acid soils, calcium ions are largely held in the colloidal clay particles. Neutral or slightly acid soils usually have a large supply of free calcium ions. Hence calcium for plant use is more available in neutral soils than in acid soils.

Testing for Reaction

All state agricultural colleges have soil testing laboratories to which farmers can send samples of soil to be tested for acidity and fertility. A farmer need only know how to take his samples properly. In addition, farmer groups have established soil testing laboratories in many counties, and students in many vocational agricultural departments test soil for acidity. Many county agents and soil conservation specialists are also equipped to make soil acidity tests. Agricultural colleges prefer to have farmers send their soil samples through their county agent. A crude test of soil

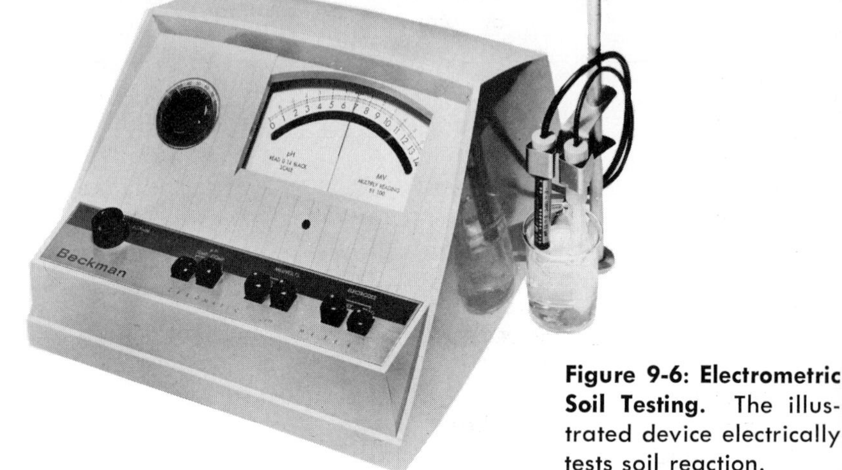

Figure 9-6: Electrometric Soil Testing. The illustrated device electrically tests soil reaction.

acidity is made by use of litmus paper. This paper when wetted and pressed on moist soil will turn red if the soil is acid and blue if it is alkaline.

Two methods for determining soil acidity are (1) an electric *p*H meter and (2) dyes which change color in relation to the *p*H of the soil.

Electrometric Method. The most accurate method of determining *p*H is by using an electric *p*H meter. When appropriate electrodes are used, an electric current is set up between the electrode and the solution which is proportional to the hydrogen ion concentration of the solution. The strength of the potential is usually read directly in *p*H units. A soil-water ratio of one to one is recommended for routine work. The suspension is stirred vigorously, allowed to stand for thirty minutes, and again well stirred just before making the measurement. For organic soils a soil-water ratio of one to five is generally used, with a standing period of two hours.

Dye Methods. Indicator dyes can be used to measure the *p*H of soils. Dye tests are accurate within half a *p*H unit, and sometimes much more closely. A sample of soil is saturated with the dye. Its color is compared with a color chart to determine its acidity. A range of *p*H from three to eight can be satisfactorily measured by the dye method.

Common colors used to indicate *p*H values by use of dyes are as follows:

Colors Indicating *p*H Values

Indicator	*pH Range*
Bromocresol green	3.8-5.6
Chlorophenol red	5.2-6.8
Bromothymol blue	6.0-7.6
Phenol red	6.8-8.4
Cresol red	7.2-8.8
Thymol blue	8.0-9.6

A number of commercial firms sell kits for testing soil acidity by use of the dye method.

Summary

Soils are either acid, neutral, or alkaline in chemical reaction. Most farmers in the eastern half of the United States are concerned with soil acidity. Soils in the western half of the United States are generally alkaline in reaction. Crops vary in the amount of soil acidity and alkalinity that they favor or tolerate. Generally crops favor a slightly acid soil. Soil acidity that crops favor is also favorable to the activities of soil microorganisms. Soil microorganisms help make plant nutrients available for plant growth. Farmers who wish to make their soil favorable for plant growth need to understand (1) the causes of soil acidity, and (2) its correction.

The major cause of soil acidity is the overabundance of hydrogen (H^+) ions attached to minute clay particles. Alkaline soils have an abundance of hydroxyl (OH^-) ions. Soil scientists express these concentrations in terms of a chemical symbol, *p*H. A *p*H of 7 designates a soil neutral in reaction. Most acid area soils range from *p*H 4 to *p*H 6.8.

Soil acidity is tested by either electrometric or dye methods and is corrected by adding lime, which supplies calcium ions displacing excess hydrogen ions. Calcium is also an important plant nutrient used abundantly by most plants that favor a slightly acid soil.

Students of soil will better understand factors which favor plant growth and soil tilth if they understand soil acidity.

Study Questions

1. Explain ionization of molecules of matter.
2. What are the causes of soil acidity and alkalinity?
3. How do *p*H symbols explain the degree of acidity and alkalinity in soils?
4. How do clay colloidal particles affect soil acidity and alkalinity?
5. How tolerant are different (1) legumes and (2) other field crops and vegetables to soil acidity?
6. How does soil acidity affect the availability of calcium for plant use?
7. How can soils be tested for acidity?

Class Activities

1. Recall fields on home farms that formerly produced good clover and alfalfa crops and now grow largely non-legume crops.
2. Make a large chart of the *p*H scale to show degrees of soil acidity and alkalinity.
3. Make laboratory tests for acidity of common products such as fresh milk, sour milk, orange juice, and lemon juice.
4. Use pieces of mica to illustrate the form of colloidal clay particles.

Liming Soils

10

Soil acidity can be corrected by adding lime materials to the soil. The amount needed will vary. The function of lime is to neutralize the hydrogen ions which cause soil acidity. In liming we are also replacing the plant nutrients calcium and magnesium which were leached from the soil or removed by crops.

Need for Lime Materials

Soil scientists estimate that United States farm lands could use about twice as much as the 25 to 30 million tons of lime materials currently applied; for some states the figure is ten times as much. Agricultural productivity could be vastly increased at a relatively low cost by such applications.

Most of our agricultural plants make their best growth on soils that are slightly acid to neutral in reaction. The purpose of using lime is to adjust the soil reaction to this favorable range. When the soil is approximately neutral, the majority of essential nutrients are near their optimum availability.

Lime is added primarily to neutralize soil acidity, although lime materials contain calcium and frequently magnesium, which are essential plant nutrients. In well-limed soils, the negative charge of clay particles and humus is neutralized by positively charged calcium (Ca^{++}) and magnesium (Mg^{++}) ions.

Lime Materials Provide Essential Nutrients. Calcium is a vital part of the cell walls of plants, so soils need the calcium supplied by lime materials for plant nutrition as well as for correcting acidity. Plants vary in the amount of calcium they require, alfalfa having particularly high requirements. Soil microorganisms, too, need calcium. Magnesium is another nutrient which may be supplied by lime materials.

Effect of Lime on Soil Tilth. Lime materials added to the soil may improve tilth, largely through the increased activity of soil bacteria and other organisms. This is especially true of soils acid enough to restrict the activity of soil microorganisms. Plant growth is stimulated by the increase in available nutrients released by soil microorganisms, and this tends to build up soil organic matter.

Lime Materials

Various types of lime products may be used, including (1) burnt lime or quicklime (calcium oxide, CaO); (2) hydrated or slaked lime (calcium hydroxide, $Ca(OH)_2$); (3) limestone or calcium carbonate ($CaCO_3$); and (4) slags, which are forms of calcium silicate ($CaSiO_3$). The above forms of lime also may appear in (1) industrial by-products, (2) marl, and (3) shell meal. *Lime materials* well describes all the products that may be used to neutralize soil acidity.

Lime materials in the forms of oxides, hydroxides, or carbonates react rather rapidly with moist, acid soils. The soil water becomes charged with calcium ions which have the ability to replace the hydrogen ions in the colloidal soil particles. The released hydrogen (H^+) ions unite with hydroxyl (OH^-) ions to form water. When the soil colloidal particles become saturated with calcium ions, the soil is no longer acid.

Calcium Equivalents of Lime Materials. Lime materials are used as a source of calcium. Consequently, the value of a liming material is dependent upon the calcium content of the material. In calcium content, 1120 pounds of "lump" or burned lime (quicklime) or 1480 pounds of fresh hydrated or water-slaked lime are equal to one ton of pure crushed limestock rock. These different weights of materials will neutralize the same amount of soil acidity.

Quick and Hydrated Lime. As indicated above, quick and hydrated lime are excellent sources of calcium for correcting soil acidity. Since the material must be prepared by processing calcium carbonate, it is usually more expensive than calcium carbonate. However, because it contains more calcium per unit of weight than ground limestone, it can be shipped for a lower charge per unit of calcium. When freight makes up a sizable proportion of the cost, quick or hydrated lime may be cheaper than ground limestone. Considerable amounts of these materials are used and they react with the soil more rapidly than ground limestone.

Crushed Limestone. When limestone, the most extensively used lime material, is finely ground, it makes a very desirable source of calcium. This is especially true if it is applied one to two months before the desired crop is to be planted. Other types of lime materials act faster than limestone but are generally more expensive.

Crushed limestone ordinarily contains both fine and coarse particles. The finely ground particles are readily available, while the larger particles serve as a source of lime for later use.

Figure 10-1: Crushed Limestone on Grass. Grass grows knee-deep on that portion of this field to which crushed limestone has been added. Note the much poorer growth, shown in the other picture, on an unlimed part of the same field.

Figure 10-2: Pasture Renovation. Heavy applications of lime materials are often needed for successful renovation of pasture. The illustrated equipment fertilizes and seeds in a single operation with minimum injury to existing sod.

Large applications of lime may supply soil needs for eight to ten years. Smaller but more frequent applications of lime are generally preferred. Clay soils require larger applications of lime than sandy or light soils, because the clay provides a greater surface for attached hydrogen ions—the cause of soil acidity.

Quality of Limestone. The fineness of crushed limestone determines its effectiveness in neutralizing soil acidity. Calcium carbonate or limestone is not easily soluble. Its solubility is increased by fine crushing. The reason for this is that the finer particles offer a larger surface to materials which can dissolve them.

The fineness of ground limestone is determined by the size of screen openings that it passes through. A 100-mesh screen has 100 openings per linear inch and 10,000 openings per square inch. The fine dust passing through a 100-mesh screen is readily available. The material passing through a 60-mesh screen is considered completely available the first year. Coarser particles passing through a 10-mesh screen will have some immediate effect, but may require several years to dissolve completely.

Larger applications of coarser materials are needed for immediate results.

Limestone usually is not pure calcium carbonate. Use of lime materials that do not have a neutralizing power equivalent to 90 to 95 per cent pure calcium carbonate is discouraged. Such impure materials must be secured at a very low price to be economical.

Most limestone contains magnesium as well as calcium. Lime material high in magnesium is derived from dolomitic limestone, $(Ca,Mg)CO_3$. This can be used to correct magnesium deficiencies in soils.

Industrial By-Products: Blast Furnace Slag. A number of industries use lime in their manufacturing processes, and after use it becomes a waste product. In localities where these industries are located, such by-products are frequently used as agricultural lime. Sources include sugar factories, paper mills, gasworks, and acetylene plants. Such lime materials may contain as much as 50 per cent water, however, which increases the cost of transportation and makes for difficult spreading. In acid areas most soil test laboratories have facilities for testing the purity of these and other materials.

Blast furnace slag is a by-product from the smelting of iron ore. Lime materials are used in this process. The slag is thus an important source of agricultural lime in steel manufacturing areas.

Shell Meal. Ground oyster shells are also used as a source of agricultural lime. They contain 90 to 95 per cent calcium carbonate.

Marl. Natural lime materials deposited in lakes, swamps, and old lake bottoms are called *marl.* Marl is generally white or gray in color, and fine enough not to require grinding, though it frequently contain shells and fragments of shells. The deposits, varying in thickness from a few inches to several feet, may be covered by soil. Marl is principally calcium carbonate, with varying quantities of magnesium carbonate and some silt, clay, and organic matter. Two cubic yards of marl are generally considered equivalent to one cubic yard of limestone. Considerable quantities of marl are used where it is available. It is generally sticky and hard to spread.

Figure 10-3: Drill Planter with Fertilizer Attachment. Lime materials should usually be applied before fertilization and planting. Earlier application of lime will help the sprouting seed make maximum use of fertilizer elements.

Time of Lime Application

Lime is usually applied before seeding a legume crop, such as alfalfa, which is sensitive to acidity. The approved practice is to apply the lime to the soil six to ten months before seeding legumes. It is a good idea to apply the lime to a previous crop such as corn, which itself may profit. The practice is especially desirable if the lime is coarse quarry screenings or granular furnace slag from which calcium is slowly available. Time and tillage operations will serve to mix the lime with the soil and to complete the neutralization process. Applying fairly generous quantities of lime even several years before alfalfa is to be planted may be helpful.

Lime applied on sod will gradually penetrate the soil and aid in maintaining a stand of acid-sensitive crops.

It may be applied on pastures and lawns at any time of the year. A fall application may be most helpful, as fall rains and the action of frost will aid the lime in penetrating the sod.

Lime materials and fertilizers should be applied separately. Lime applied before fertilizers will tend to make some fertilizer elements more readily available for plant use.

Lime materials left on the surface of fields subject to erosion can be readily lost during heavy rains. When they are mixed with the soil, the only lime removed would be that contained in the eroded soil.

A soil-borne disease such as potato scab is encouraged by lime. Therefore, in a rotation which includes potatoes, lime should be applied after the potato crop.

Method of Application

If acidity is corrected only in the surface soil, deep-rooted legumes such as alfalfa will not do well. Lime placed on the surface soil and plowed under will be buried beneath the surface. The surface soil in this case may remain acid. Lime materials need to be worked well into the soil. This can be done in preparation of the seedbed. Plowing under one-half the application of lime, and disking- and harrowing-in the remainder, will provide a good distribution. In spreading lime on pastures, it can be worked in by disking.

Present-day commercial operators are usually well-equipped to spread lime from trucks. They can spread at varying rates in different parts of the field, as needed. Soil tests will indicate areas needing heavier applications. Such spots can be indicated by stakes on the field and marked on a map of the field.

A manure spreader can be used for spreading lime if the lime is put on six to eight inches of manure or straw in the bottom of the spreader. Another procedure is to tack a canvas over the manure spreader's apron slats to prevent the lime materials from sifting through.

Figure 10-4: Endgate Spreader. Lime materials are often applied with equipment such as this.

Figure 10-5: Measured Application. The illustrated boom spreader can be adjusted to supply lime materials evenly and in recommended amounts.

Two other types of lime spreaders are (1) a two-wheel, box type of spreader, and (2) an endgate spreader. Both are satisfactory. The endgate spreader is more dusty than other means; also, the lime materials need to be shoveled into it.

When lime is spread from heavy equipment such as trucks, there is some danger of packing the soil too tightly. For this reason, it is best to wait until the soil is fairly dry before spreading lime by this method on cultivated fields. Sod is more resistant to packing.

Amount of Lime to Apply

The amount of lime needed per acre depends upon (1) the degree of acidity of the soil, (2) the buffer capacity of the soil, (3) the acidity of the subsoil, (4) the crops to be grown, (5) the grade or purity of the lime materials, and (6) the frequency of application. For most crops, a soil *p*H of 6.0 to 6.8 is most desirable. Alfalfa favors a *p*H of 6.8. Potatoes are grown in soil with a *p*H of 5.0 to 5.2 to control scab.

State agricultural colleges usually make recommendations on the amount of lime materials needed for soils in their states. It is a good practice for farmers to find out what amount their agricultural college recommends. Sometimes, more frequent applications of lime are better than one large application.

Where soils have not been limed prior to the planting of a legume crop, 500 pounds of 10-mesh limestone per acre can be drilled into the soil, using the fertilizer attachment of the grain drill. This lime application will take care only of the current crop needs.

A general recommendation for amounts of lime needed to correct usual types of soil acidity is as follows:

Tons of Lime Required to Correct Usual Types of Soil Acidity

Soil Texture	*Slightly Acid*	*Moderately Acid*	*Strongly Acid*
Sand and loamy sand	¼-1 ton	1-2 tons	2-3 tons
Loam and silt loam	½-2 tons	2-3 tons	3-6 tons
Silt and clay loam	½-3 tons	3-6 tons	6 or above

Over-Liming. Some danger exists in too-heavy applications of lime. Crops vary in the degree of acidity they favor. Light sands may easily be over-limed. More frequent and lighter applications of lime materials may be the most desirable practice.

Summary

Soil acidity can be corrected by adding lime materials to the soil. In liming we are also replacing the plant nutrients calcium and magnesium which were leached from the soil or removed by crops. Scientists estimate that American farmers need to use twice as much lime as they now do; in some states, farmers need to use ten times as much.

Lime materials include burnt lime or quicklime, hydrated or slaked lime, ground limestone, slags, other industrial by-products, marl, and shell meal. The quality of lime materials should be tested before application.

The time when application will be most beneficial varies for different crops. Methods and machinery for application also vary, though many farmers find it most satisfactory to hire commercial operators with their own trucks for the purpose of spreading lime materials. The amount of lime needed per acre depends upon (1) the degree of acidity of the soil, (2) the buffer capacity of the soil, (3) the acidity of the subsoil, (4) the crops to be grown, (5) the quality or purity of the lime materials, and (6) the frequency of application. There is some danger in over-application of lime.

Study Questions

1. How much more lime should farmers use than they are using at present?
2. What choice of lime materials does a farmer have?
3. What are factors to consider in use of limestone for correcting soil acidity?
4. How usable are lime materials such as slag and marl?
5. When should the magnesium type of limestone be used?
6. When should lime be applied?
7. What are acceptable methods of lime application?
8. How much lime should be applied?

Class Activities

1. Test soils from home farms for acidity with testing kits.
2. Take a class field trip to practice taking soil samples and mapping a field for acidity testing.
3. Take a field trip to a soil testing laboratory, if one is located nearby.
4. Demonstrate appliction of lime where needed on a class member's farm.
5. Make a display of different kinds of lime materials.

Alkali Soils

In humid regions, the soluble materials resulting from soil formation are leached from the soil by rainfall and carried away into creeks and rivers. These salts may eventually reach the oceans, where strong salt solutions have been accumulating for millions of years. Similarly, lakes that do not have outlets become salty, because the salt from inflowing water accumulates while much of the water itself is lost through evaporation. Great Salt Lake in Utah and the Dead Sea in Palestine have become salty in this way.

In dry areas, there is not enough moisture moving through the soil to remove the salts, and they remain. The soils of the Western states frequently contain an excess of salts, and are technically called *salty soils*. Through a change in conditions, other soils that once were salty have developed into *sodic soils*. Both soil conditions are indicated by the term *alkali*. How to recognize and correct them is a problem confronting many farmers and ranchers.

The Two Types of Alkali Soils

Our knowledge of alkali soils has increased tremendously since 1900. With our better understanding of these soils it has been necessary to introduce new terms. We now know that there are two distinctly different kinds of alkali soils.

Salty Soils. Salty soils, the first of the two types, are soils in which there has been an accumulation of soluble salts—table salt ($NaCl$) is only one of many salts recognized by the chemist.

Among the salts which are prevalent in these soils are sodium sulfate (Na_2SO_4), magnesium sulfate ($MgSO_4$), magnesium chloride ($MgCl_2$), sodium chloride or table salt ($NaCl$), and sodium bicarbonate or baking soda ($NaHCO_3$). Usually sodium is the most abundant positive ion. Soils which have an accumulation of such salts have been called "alkali soils," "white alkali soils," "saline soils," and "salty soils." *Salty soils* is now the preferred term. Salty soils are usually well aggregated and have good physical condition, but the abundance of salts in the soil solution makes plant growth difficult.

Sodic Soils. The second type of alkali soil is formed in a manner which is somewhat related to the formation of acid soil, and a review of Chapter 9, "Soil Reaction," may aid in understanding it. These soils occur in areas where salty soils were once found. However, conditions have changed so that most of the salts have been leached out. During the leaching, sodium has replaced much of the calcium which is normally attached to the colloidal

Figure 11-1: Salty Soil. Nevada agricultural experts inspect a salt spot in a field. In this case, two or three heavy irrigations will correct the condition.

soil particles. When this happens in the absence of abundant salts, the soil particles are no longer strongly attracted to each other. Thus, the soil aggregates break down and each soil particle becomes independent of the others, so that the soil is said to be *dispersed.* The large pores between soil aggregates are destroyed and water cannot move easily through the soil. The soil is extremely hard when dry and sticky when wet, so it is difficult to cultivate. Such soils have been known as "alkali soils," "black alkali soils," and "sodic soils." *Sodic soils* is now the accepted term.

The term *alkali* is still used to indicate both groups of soils.

Salty Soils

Why Salty Soils Form. In the weathering of parent material to form soils, many soluble materials are formed. When rainfall is sufficient, these soluble materials are leached out. When all the rainfall is either used by plants or evaporated from the soil surface, water does not pass through the soil, and the soluble materials remain.

Even in dry areas, the salts formed from normal parent materials are not usually abundant enough to cause serious difficulty. However, there is a tendency for salts to move with water to low spots, where temporary lakes form after rain. The salts from large areas of land may move to these low spots and become abundant enough to cause trouble.

When soils in a dry area are put under irrigation, salt problems usually develop. Irrigation water dissolves salts that were built up during soil formation, and they move with the water. Under irrigation, the water table is usually raised because some of the applied water passes through the root zone and takes salts with it. When the water table becomes high enough, water and salts move to the surface by capillarity. The water evaporates and leaves the salts to form a salty soil. This happens most frequently in low areas or at the foot of a slope, such as a river terrace. All river water used for irrigation contains some salts and these salts remain in the soil after the water has been used by plants. River waters that are especially high in salts increase the problem.

The build-up of salts in soils following irrigation is believed to have caused some of the ancient civilizations to decline. Our present knowledge of chemistry makes it possible to devise methods of combating the problem.

How Salts Injure Crops. In salty soils, the soil solution tends to retain its water rather than give it up to plants. Much of the soil water enters the plant root by *osmosis*. This, you will recall, is a process by which water tends to move through a semi-permeable membrane from a dilute to a more concentrated solution. Frequently, plant nutrients are carried along with the water.

In ordinary soils, the root hairs contain a more dense solution of salts and sugar than the surrounding soil water and thus absorb the soil water and its nutrients. In salty soil, the soil solution may be almost as dense as the root hair solution. In this situation the root hairs will fail to absorb adequate soil moisture. Plants may not be killed, but will fail to make normal growth. Seeds planted in strongly salty soils may fail to germinate because they are unable to absorb soil moisture.

Figure 11-2: Border Dike Irrigation. Unless provision is made for drainage, salt accumulation may become a problem in irrigated fields.

Figure 11-3: Drainage Ditch. This Western farmer prevents salt accumulation with an adequate provision for draining excess irrigation water from his fields.

Testing Salty Soils. Because a salt solution conducts an electric current, the prevalence of salts is usually measured by the ability of the soil solution to conduct a current. When conductivity is high, we know that the soil contains excessive amounts of salt. The salts can also be analyzed chemically.

The composition of a sample of salty soil found in Montana is shown by the following analysis.*

Composition of Salty Soil in Montana

Component	*Per Cent*
Silica (soluble)	0.11
Calcium sulfate	4.84
Magnesium sulfate	23.83
Sodium sulfate	59.67
Sodium chloride (common salt)	7.86
Sodium carbonate (sal soda)	3.13
Potassium sulfate	0.54

Correction of Salty Soils. A very effective way to remove salts is to flood the land with irrigation water and wash them out of the soil. This requires an effective drainage system. An occasional

* Burke, Edmund, *Alkali Soils and Waters in Montana,* Circular 101, Montana Agricultural Experiment Station, January 1939, p. 11.

heavy application of irrigation water is also an approved practice to prevent an accumulation of salts. It should be noted again that salts will move into low areas with drainage water. The drainage system must be designed to intercept this water before the problem can be corrected on lower land.

Improving the fertility of a soil will make plants more tolerant of salty conditions. Thus a liberal application of barnyard manure or the use of a green manure crop frequently tends to overcome mild saline conditions. These treatments also tend to improve soil structure. The improved structure allows water to pass more readily through the soil and encourages leaching.

Plants are most sensitive to injury from saline conditions in the seedling stage. It is very desirable to keep the soil moist so that the soil solution will be as dilute as possible during this stage of growth.

Crops Tolerant of Salty Soils. Under some conditions, farmers find it difficlt to remove the excess salt from soil. In such cases, the best practice is to grow crops that are tolerant of salts.

Tolerance of Fruit Crops to Salty Soil

High Salt Tolerance	*Medium Salt Tolerance*	*Low Salt Tolerance*	
Date palm	Fig	Pear	Almond
	Olive	Apple	Apricot
	Grape	Orange	Peach
	Cantaloup	Grapefruit	Strawberry
		Prune	Lemon
		Plum	Avocado

Tolerance of Field Crops to Salty Soil

High Salt Tolerance	*Medium Salt Tolerance*		*Low Salt Tolerance*
Barley (grain)	Rye (grain)	Corn (field)	Field bean
Sugar beet	Wheat (grain)	Flax	
Rape	Oats (grain)	Sunflower	
Cotton	Rice	Castor bean	
	Sorghum (grain)		

Tolerance of Vegetable Crops to Salty Soil

High Salt Tolerance	*Medium Salt Tolerance*		*Low Salt Tolerance*
Garden beet	Tomato	Potato	Radish
Kale	Broccoli	Carrot	Celery
Asparagus	Cabbage	Onion	Green bean
Spinach	Bell pepper	Pea	
	Cauliflower	Squash	
	Lettuce	Cucumber	
	Sweet corn		

Tolerance of Forage Crops to Salty Soil

High Salt Tolerance	*Medium Salt Tolerance*		*Low Salt Tolerance*
Saltgrass	White sweet clover	Tall fescue	White Dutch clover
Nuttall alkaligrass	Yellow sweet clover	Wheat (hay)	Meadow foxtail
Bermuda grass	Perennial ryegrass	Rye (hay)	Alsike clover
Rhodes grass	Mountain brome	Oats (hay)	Red clover
Rescue grass	Strawberry clover	Orchardgrass	Ladino clover
Canada wildrye	Dallis grass	Blue grama	Burnet
Western wheatgrass	Hubam clover	Meadow fescue	
Barley (hay)	Alfalfa	Reed canary	
Bird's-foot trefoil		Smooth brome	
		Tall meadow oatgrass	

Sodic Soils

So long as the concentration of soluble salts in soil remains high, the soil tilth is usually satisfactory. These soluble salts also prevent the soil from becoming excessively alkaline. When conditions change so that most of the soluble salts are leached or otherwise removed, the soil properties are altered materially. In *sodic soils* a considerable portion of the calcium which is normally attached to the soil particles has been replaced with sodium from the salts. When the soluble salts are removed, the soil particles with sodium attached dominate the soil properties. Sodium does not neutralize the negative charge on colloidal soil particles as well as calcium or hydrogen does. The particles retain part of their negative charge and repel each other, causing the soil structure to break down. The soil becomes dense and will not take water readily. It frequently becomes so alkaline that organic matter becomes soluble. This causes the soil to become dark in color. The dark color taken on by these soils is the basis for the expression "black alkali."

Figure 11-4: Adding Gypsum to Irrigation Water. Sodic soils are often corrected with gypsum.

Correction of Sodic Soils. To correct a sodic soil condition, i is necessary to replace the attached sodium with calcium. Th reclamation process is somewhat related to that of correctin soil acidity. However, in sodic soils, calcium carbonate (lime does not supply enough soluble calcium to replace the sodiu in a reasonably short time. Because of this, lime alone is not a effective treatment. The problem is to get a source of solubl calcium. Gypsum, which is hydrous calcium sulfate, is usually the cheapest source. When gypsum is needed, the rate of application has to be quite high. It usually ranges from two to 20 tons per acre. Gypsum can be spread on the soil surface like lime and leached into the soil with water. It can also be dissolved and applied in irrigation water.

Because sodic soils frequently contain calcium carbonate (lime), they can also be corrected by increasing the solubility of the calcium already in the soil. This can be done with a number of compounds, of which sulfur is the most important. Sulfur

Figure 11-5: Truck Spreader. Trucks such as this can be used to spread gypsum and sulfur as well as lime and fertilizer. Many farmers find that application of gypsum by this method is more satisfactory than application in irrigation water.

is oxidized to sulfuric acid by microorganisms in the soil. The sulfuric acid then reacts with the calcium carbonate of the soil to form gypsum. In effect, then, when the soil contains lime, the use of sulfur brings about the same corrective action as would the use of gypsum in lime-deficient soil.

One pound of sulfur produces about five pounds of gypsum by reacting with soil lime. Thus sulfur can be applied at about one fifth the rate of gypsum. After application, sulfur should be mixed through the soil by cultivation.

After calcium becomes attached to soil particles, the sodium remains as sodium sulfate. This can be leached out of the soil by rain or irrigation water.

Salty-Sodic Soils

Some soils have characteristics of both the salty and sodic types. They contain enough exchangeable sodium to interfere with water movement and enough soluble salts to restrict plant growth. The structure of these soils is usually so poor that a source of soluble calcium must be applied before the salts can be leached from them.

Demonstrating Salty and Sodic Soils

First Step. Place about 100 grams of normal soil on a Buechner funnel. Leach with 200 cc. of distilled water under suction. Record the time required for the last 50 cc. of water to pass through the soil. Test the reaction of the water before and after leaching.

Second Step. Leach the soil with 300 cc. of NaCl salt solution under suction; use 58 grams NaCl per liter. Test the reaction of the NaCl solution before and after it has passed through the soil. Record the length of time required for the last 50 cc. of NaCl solution to pass through the soil. You now have a salty or white alkali soil.

Third Step. Leach the soil with 200 cc. of distilled water under suction or until the solution becomes dark-colored because of dissolved organic matter. Test the reaction of the last water to come through. You now have a sodic or black alkali soil.

Explanation. You start with a normal soil. The rate of water movement through it is a measure of the soil properties and texture. The *p*H of the water after passing through the soil will be approximately that of the soil.

Leaching with a NaCl solution converts the soil to a sodium saturated soil similar to that in salty soils. Note that the presence of excess $NaCl$ prevents structural breakdown or a pronounced increase in alkalinity.

On leaching a second time with water, the free $NaCl$ is removed and sodium has replaced the calcium normally attached to the soil particles. You now have sodic or black alkali soil. The last water extract to come through is usually dark-colored, indicating that the organic matter has been dispersed and removed in the water. This extract should have a *p*H above 8.5.

Summary

In dry areas, there is not enough moisture moving through the soil to remove the salts resulting from soil formation, and *salty soils* result. Irrigation can increase the problem of soil salts if drainage is inadequate. Salty soils injure crops because they tend to prevent roots from absorbing sufficient moisture and nutrients, and to prevent seed from germinating. Salty soils may be corrected by flooding the excess soluble salts out of the soil with irrigation water. When this is done, it is necessary to provide adequate drainage in low-lying areas. Improving soil fertility will make plants more tolerant of salty conditions. When excess salts cannot be removed from soil, it is best to choose salt-tolerant crops.

When conditions change so that most soluble salts are leached or otherwise removed, *sodic soils* result. These are soils in which the particles with sodium attached dominate the soil properties. The soil structure breaks down; the soil becomes dense and will not take water readily, and organic matter dissolves. The condition is corrected by replacing the attached sodium with calcium. Gypsum (hydrous calcium sulfate) is often used for this purpose. It can be spread on the surface like lime and leached into the soil with irrigation water, or dissolved into irrigation water. When sodic soils contain lime, they may be corrected with the application of sulfur.

Both salty and sodic soils are called *alkali soils.*

Salty-sodic soils have characteristics of both types. To correct them, soluble calcium usually must be applied before the salts can be leached out.

Study Questions

1. What kinds of alkali soils are there?
2. What effect does an excess of attached sodium have on soils?
3. How does an excess concentration of soil salts affect plant growth?
4. What are the characteristics of a salty soil?
5. How can salty soils be corrected?
6. What are the characteristics of sodic soil?
7. How can sodic soils be corrected?
8. How can one recognize and correct soils with both salty and sodic characteristics?
9. How does sulfur correct sodic soils?

Class Activities

1. Test sodic soils for available calcium by pouring a teaspoonful of sulfuric or hydrochloric acid into a cup of soil. If it bubbles, it will indicate presence of lime.
2. Find soils of the community that are classed as alkali and discuss their condition.
3. Visit and observe alkali soils and determine their tilth and water-absorption abilities.
4. Treat black alkali soils with soluble calcium materials and observe their improvement.
5. Arrange to visit farms in which soils are being treated for alkali conditions.

Testing for Soil Fertility

12

Soils differ tremendously in their ability to supply nutrients to plants. When a soil is low in its ability to provide plants with one or more nutrients, it is usually profitable to add these nutrients to the soil. American farmers are now using about twenty million tons of fertilizer a year. The main plant nutrients in fertilizers are *nitrogen, phosphorus,* and *potassium.* When a farmer invests money in fertilizer, he expects to pay the cost out of the increased yield from his crop. He is interested in using fertilizer on only those acres where yields will be increased. When yield or crop quality has not been improved from the use of the fertilizer, the farmer has made a poor investment.

Applying commercial fertilizer without a soil test is like buying a box at an auction without knowing what is in it. Fertilizers are expensive. Hundreds of dollars can be saved by farmers who know what mixture to buy.

The dairyman buys cows and bulls according to production records. Beef producers practice bull indexing, a method of recording the rate of growth of young bulls. Sheep producers pay attention to such factors as frequency of twins and pounds of wool per animal. Similar indices on soils are available to farmers and ranchers through testing. Soils can be modified more easily than livestock.

Soil scientists and commercial companies have made tremendous progress in developing accurate and reliable soil tests. Results from laboratory tests can be checked by the individual farmer with fertilizer trials on growing crops.

What Are Soil Fertility Tests?

The major purpose of soil tests is to estimate the ability of soil to supply the various nutrients under expected conditions. Plant tissue tests of growing leaves and stems can also aid in determining whether plants are obtaining the nutrients they need.

During the last 30 years much progress has been made in improving the accuracy of soil testing. Many farmers now base their fertilizer use on such tests, since the cost is small enough to be covered by a saving of a few pounds of fertilizer. Soil testing has been so widely accepted that over one million soil samples are analyzed in the United States each year. Soil tests have the advantage that they can be made before seeding a crop. Fertilizer can then be applied to the soil at the appropriate rate and at the convenience of the farm operator.

Accuracy of Soil Tests. A chemical soil test attempts to measure how much of a particular nutrient the soil will provide to a crop during its growth period. In soil tests it is assumed that the test can extract nutrients in rough proportion to the amount the crop can remove. Because of the extreme complexity of both plants and soils, completely reliable predictions of the availability of a

Figure 12-1: Soil Testing in the Laboratory. Modern laboratories such as this ensure greater reliability for soil tests.

Figure 12-2: Fertilization and Moisture. When choosing their fertilizers, farmers should take into consideration such field conditions as availability of moisture. A well-watered crop can profitably use more fertilizer than the same crop grown under droughty conditions.

soil nutrient cannot be expected.

It must be kept in mind, too, that some soil tests may be more accurate than others. Thus a soil test should not be used until information is available on its reliability in the area. Extensive field and laboratory research data are necessary to determine reliability.

A farmer may have had considerable past experience with fertilizer on his land. This aids his judgment on how much increase in yield he can expect from the use of fertilizer on a given crop. The experience his neighbors have had on similar soils is also helpful. He can usually secure specialized help from the local vocational agriculture department, the county agent, or the Soil Conservation Service.

The weather pattern to which the crop will be subjected and the prevalence of weeds, plant diseases, and insect pests enter into the farmer's chance of securing a profitable return from his fertilizer investment. He is concerned not only with which nutrients to apply, but also with which rate of application will provide the best return.

Tests for Phosphorus. Major emphasis has been placed on devising a test for phosphorus. Various types of tests are now available. In many locations these tests can be used to make fairly reliable fertilizer recommendations. Several *extracting solutions* are in use, including distilled water, weak and strong acids, and the salts of weak or strong acids. Usually, after the extraction has been made, the solution develops a color which reflects its phosphorus level. A simple comparison of the intensity of the color in the solution with standard color charts is frequently used, but delicate instruments for making such measurements are also available.

Tests for Potassium. Tests for determining the available potassium level in soils have also received considerable attention. Many of these tests are based on the amount of potassium that is present in exchangeable form. Most soils are much higher in total potassium than in most other essential nutrient elements. Much of this potassium is present in forms that plants cannot readily extract.

Tests for Nitrogen. Most soil nitrogen is a part of the soil organic matter. As the organic matter breaks down, nitrogen is released for use by plants. The nature of the organic matter present, the temperature and moisture of the soil, and the availability of other nutrients—all influence the rate of nitrogen release. Some progress in predicting the nitrogen-supplying power of a soil is being made by studying the rate at which

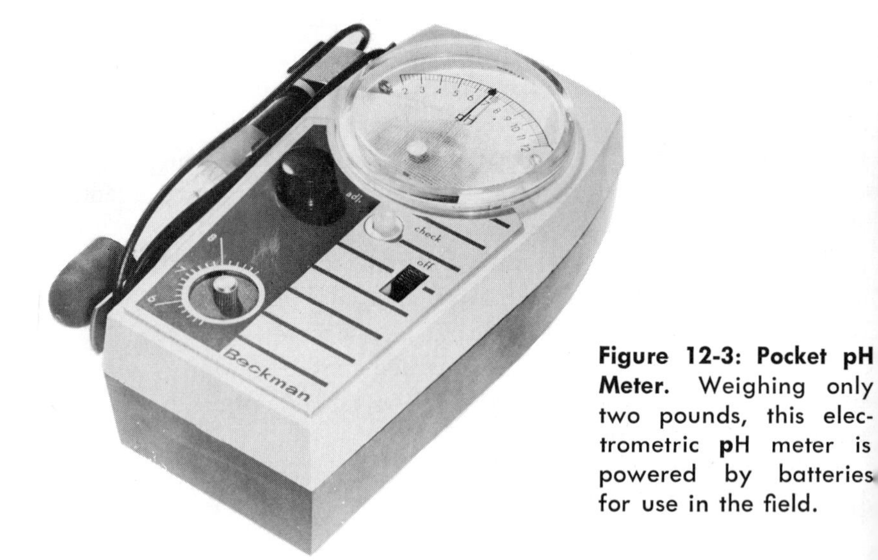

Figure 12-3: Pocket pH Meter. Weighing only two pounds, this electrometric pH meter is powered by batteries for use in the field.

nitrates are produced in a soil incubated under standard temperature and moisture conditions. The amount of organic matter oxidized by a weak oxidizing agent also seems to provide some measure of the rate at which nitrogen is released to plants.

Tests for Acidity and Alkalinity. Most soil tests include a determination of acidity. This can be done with appropriate indicator solutions. In most soil testing laboratories, however, acidity is determined with electrometric equipment. The acidity of a soil is an extremely important characteristic. Knowledge of this factor aids in predicting the availability of plant nutrients.

Laboratory Soil Tests

In most states, some form of soil testing service is available. A charge is made for the analysis. In some states, testing laboratories are available in many of the centrally located cities of a county. Usually the county agent or vocational agriculture teacher will be familiar with the soil testing services available and have information about the reliability of the tests. Considerable use is made of commercially manufactured soil testing kits by individual farmers and by high school vocational agriculture departments. Work of established soil testing services is considered to be most reliable.

Taking Soil Samples. Perhaps the most important single job a farmer has, is to send to the laboratory accurate, representative samples of his fields. The size of the sampled area depends upon the uniformity of the soil. Samples are taken from various parts of a field to make up a single representative sample. They should be taken from 15 to 20 locations in a given area. These 15 to 20 sub-samples are thoroughly mixed together, and a portion of the mixture then is used to fill a sample box holding approximately one pint. The sample should be numbered as soon as it is taken, and should be dried at room temperature. Heat used in drying may lead to inaccurate results.

When preparing to take a soil sample, scrape away surface litter. Samples are usually taken from the plow-layer or the first six inches of surface soil.

Avoid areas of the field which are not representative. Nonrepresentative areas are: (1) dead furrows, (2) back furrows,

Figure 12-4: Soil-Sampling Equipment. A clean trowel, soil auger, or shovel may be used for taking soil samples. A clean pail will serve for collection and mixing. Soil containers and a large mailing tube are at the right of the pail. Note the paper and pencil for recording the samples. The other picture shows a soil auger in action. The soil auger is especially convenient to use for sampling pasture and lawn.

Figure 12-5: An Area to Avoid. Spots where lime or fertilizer has been applied in previous years should never be chosen as sites for taking a soil subsample.

(3) old fence lines, (4) eroded spots, (5) old haystack bottoms, (6) field terraces, (7) field depressions, and (8) alkali spots. These non-representative areas can be sampled by themselves, if separate sampling is justified.

In taking soil samples from cultivated fields, avoid rows that have been fertilized. The samples should be taken between rows.

A good management practice is to sample a field for soil testing once in a complete crop rotation. Take the samples in the early fall while the field is still in sod. Fertilizer applied on a cultivated crop usually means a quicker return from the money invested. However, if lime is needed for a legume crop, it should be applied up to a year before, to allow time for the lime to correct soil acidity. Fall sampling and testing allow the farmer plenty of time to buy the appropriate fertilizer and make plans for spring application.

In a recently plowed field, a handful of surface soil from each of 15 to 20 places may be used. On fields that are not freshly plowed, the sample should be taken from the surface down to a depth of six or seven inches. On pasture land which has not been plowed, the sample should be taken from a depth of three to five inches.

A shovel, spade, post-hole digger, or soil auger may be used. A clean 10-12 quart bucket or pail will serve to collect and mix the individual samples.

If a six-inch or seven-inch hole is dug, take a one-inch slice of soil along the face from the top to the bottom of the hole. Special soil sampling tubes can be purchased. They have a split-side to ease removal.

Subsoil Samples. Farmers producing deep-rooted crops such as alfalfa and trees need to take subsoil samples. These should be taken at a depth of 12 to 18 inches. An auger or sampling tube facilitates this. If neither is available, a 12-inch hole may be dug. The soil at the bottom of the hole should be loosened to a depth of 18 inches. A handful of the loosened soil will serve as a sample. Approximately five handfuls from various places are needed to make a representative subsoil sample.

Recording Soil Samples. Soil testing laboratories usually provide a sampling information sheet as well as soil sample mailing

Figure 12-6: Taking the Sample. First, a shovelful of soil is removed from a hole six to eight inches deep. In the second picture, a uniform vertical slice of soil from the side of the hole is being put into the clean pail. In the third picture, the soil from 15 such subsamples is being carefully mixed. Finally, a small portion of the mixed, representative sample is being put into a container.

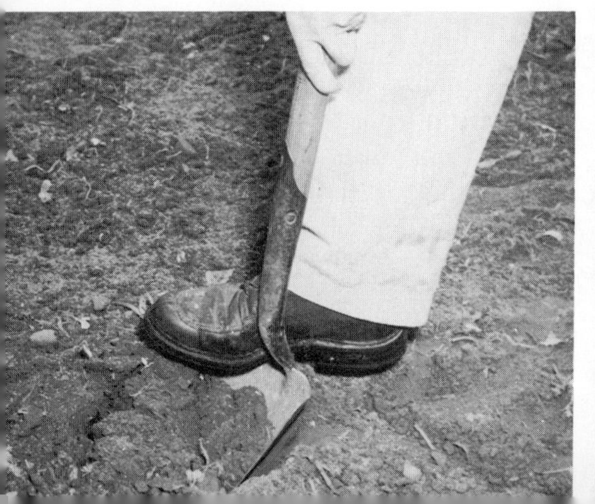

cartons. However, if a farmer is making his own test he needs to identify the field location of his soil samples.

A map should be made of the fields sampled to show the area represented by each soil sample. The farmer should keep a copy for himself. To assist the soil testing specialists in locating the area, the standard map of the fields sampled should show the section, township, and range. A cropping history should indicate crops grown for the past four or five years and their yield. A record of past yields aids in isolating special soil problems. Other questions to be answered include: Was a legume crop plowed under? Was summer fallow used? When were lime and fertilizer last used and at what rate per acre? Will the crop be grazed or harvested? Will irrigation be used? Will the fertilizer be applied by drill, planter, or other means? Reports of previous soils tests should be noted, and so should the crops to be grown the next year.

The above information will aid both the soil testing experts and the farmer in determining the amount and kind of fertilizer, lime, or other soil amendments to use.

How Laboratories Test Soil. Soil testing laboratories are usually set up to make large numbers of tests at one time. The soil samples are first ground with a mechanical grinder, and then thoroughly mixed so a small sub-sample is representative of the entire sample and the farmer's field.

The actual sample for analysis is usually measured by using a container which, when leveled, holds the appropriate amount of soil. Extracting solutions are usually added with automatic pipettes. If one extracting solution can be used to test for several

Figure 12-7: Recorded Samples. These soil samples are numbered and identified by the field from which each was taken.

nutrients, the efficiency with which tests can be made is greatly increased. Mechanical agitation is generally used to mix the extracting solution and the soil sample. The extraction flasks are fitted into racks by means of which ten to twenty flasks are agitated at the same time. When filtering is used, the filters are arranged so that a group of extracts can be poured at one time from several extraction flasks through the filter papers. Systems for keeping records and sending out results are developed to a degree that this phase of the work can be handled efficiently and accurately.

Soil Acidity and Alkalinity Tests. Soil reaction (acidity or alkalinity) is indicated by the *p*H symbol. Generally speaking, nutrients are most readily available in soils which are slightly acid to neutral—*p*H 6 to *p*H 7. The soil testing specialists know how to adjust fertilizer recommendations to offset adverse acidity or alkalinity levels.

Acidity and alkalinity tests also are relatively easy to make with commercial soil testing kits. A farmer or gardener who knows the acidity and alkalinity of his soil can better estimate the fertilizer needs of his plants.

Acid soils are generally corrected by use of lime materials, and alkaline soils by sulfur materials. Lime is the common source of calcium, which in itself is an important plant nutrient. Soil experts make a point of the widespread need for the most extensive use of lime materials for acid soils. Many fertility problems can be partially solved by soil treatment for acidity or alkalinity.

Interpreting Soil Tests. Many factors are taken into consideration in interpreting soil tests. Among these are the level of plant nutrients established by the tests, and the fertilizer response in the area. Some soil types are more likely to respond to certain fertilizers than other soil types. Thus good soil maps aid in interpretation of soil tests. The plant-nutrient needs of the crop to be grown are always given consideration. Information on the yield that has been secured from different crops grown on the land for the past few years is helpful. If manure or fertilizer has been applied to the land during the past few years, this information should be available. If the land has recently grown a legume crop, responses to nitrogen fertilizer are less probable.

Agricultural colleges and testing laboratories generally send their test results to the county agent. He in turn is expected to assist the farmer or rancher in determining the best fertilizer practices to follow.

Soil Testing Kits

Soil testing kits are extensively used by farmers, county agents, and vocational agriculture classes. Several companies offer them for sale. They are generally put up in a handy box or kit containing testing apparatus, chemical reagents, and directions for reading test results.

Such kits allow farmers to make spot checks of their soils during the growing season. If the farmer observes deficiencies in his growing crops, he can attempt to make improvements based upon tests for plant nutrient shortages. Certain spots in a field or

Figure 12-8: Soil Testing Kit. Here being used by an expert, the portable soil testing kit helps many farmers grow better crops.

certain fields may show nutrient lacks more than others. However, the farmer must keep in mind that poor crop conditions may be due to such factors as poor drainage, lack of moisture, acid or alkali conditions, plant diseases, and insects.

From soil testing kits used at schools, boys can get first-hand knowledge of the methods and the number of problems involved. A laboratory experience in soil testing, plus field experience in taking soil samples and interpreting results, do much to develop interest in soils and soil management.

Reliability of Soil Testing Kits. Persons using commercial soil testing kits should attempt to find out their adaptability to their general area. Some states have conducted such tests. Soil testing kits manufactured in the state where used are usually quite reliable for that state. Accuracy of any kit-test can be checked by sending identical soil samples to an official soil testing laboratory to see if the results are the same.

Precautions in Using Soil Testing Kits. Soil testing kits become more valuable to the farmer as he learns more about his soils and crops. To a beginner, however, results from soil testing kits can be misleading. Precautions to be observed are:

1. Use care in securing representative soil samples.
2. Check the reagents (acids, and so on) in the soil kit—some are sensitive to light and air, and lose their power.
3. Use clean glassware and other laboratory aids.
4. Use care in reading results.
5. In interpreting results, consider field conditions and other factors affecting crop growth.

Plant Tissue Tests

Growing plants can be chemically tested to determine the amount of nutrients they contain. Plant tests enable the grower or soil scientist to look inside the plant, so to speak. The testing process involves a chemical analysis of a portion of the plant to determine the concentration of nutrients in such degrees as low, adequate, and plentiful.

Tissue tests have an advantage over soil tests because the plant itself is checked as a nutrient-extracting agent. A plant cannot build new tissue unless it has an adequate amount of each nutrient. The plant builds itself a structure much as a contractor builds a house. Unless the proper materials are available at the proper time, the building program is stopped. The rate of plant growth is controlled by the one essential nutrient present in the smallest amount. If tissue tests show a nutrient to be present in amounts considerably above the minimum required for growth, the analysis indicates that the nutrient is adequately present. If one nutrient element is deficient, there is a tendency for other nutrients to accumulate in the plant. The analysis reflects the condition of the plant at the instant the sample was taken. Changes in nutrient status of soils and changes in the stage of growth of the plant can occur from day to day.

Better than a soil test can be expected to do, plant tissue tests reflect the conditions a plant is encountering. Before plant tissue tests can be made, however, the crop has to be established. With many crops, the information from tissue tests is secured too late in the season to be very helpful in correcting the problem for that crop. Corrective treatment then can be made only for succeeding crops.

Kinds of Plant Tissue Tests. There are two types of plant tissue tests. The first of these is referred to as a *green-tissue test*. Green-tissue tests are made in the field from growing crops. The second type is known as *plant analysis*. This is made in laboratories.

Green-Tissue Tests. In making a green-tissue test, you should secure a representative sample of tissue from plants with growth

Figure 12-9: Making Green-Tissue Tests. Field kits show how well growing crops are extracting nutrients from the soil. Can you identify the two kinds of tissue samples that are being made?

characteristics similar to those which prevail over the field. If this precaution is taken, usually a sampling of eight to twelve plants is adequate to represent the field. Diseased or abnormal plants should be avoided. The part of the plant from which the sample is taken is also important. Some nutrients are readily translocated from one part of the plant to another, and the plant usually translocates these nutrients to new tissue even at the expense of reducing plant nutrients in older tissue to a critical level. Thus the lower leaves on a corn plant may show nitrogen deficiency while the plant is still making new growth. Some other nutrients are not readily translocated to new tissue, and under these circumstances new tissue is most likely to suffer a deficiency. Thus the part of the plant on which analysis is made influences the value of the test.

The plant has its greatest requirement for nutrients when it is making the most rapid growth. This is usually near flowering time. The most deficient nutrient usually controls the rate of growth, and all others will tend to accumulate. Thus the test will usually show only one nutrient to be at a critical level. If the status of this nutrient is changed, another may quickly become critical. If plants are growing rapidly, a good nutritional balance is indicated. Under these circumstances all nutrients are rapidly being built into plant material, and the tests may show lower levels for the various nutrients than are found in slower-growing plants—except in the case of the critical nutrient.

Making Green-Tissue Tests. Green-tissue testing field kits can be purchased. Directions are provided in the kit.

The first step is to make thin slices of plant tissues. Chemicals then are added to the exposed tissues and plant juices. Color changes in the chemicals indicate the amount of plant nutrients present.

Slices of tissues are made from different parts of different types of plant. With corn plants, slices are made on the stock. With broad-leaf plants, slices are made of the petiole, the slender stem that supports the leaf. With grasses or grains and other leaf plants, sections of the leaf are sliced thin with a razor blade.

Farm operators should watch plant growth for hunger signs. It is a good practice to test tissues in healthy fields and to compare results with tests on less-healthy plants.

Figure 12-10: Field Trial. Field trials on the farm provide the most accurate fertilizer tests.

Fertilizer Field Trials

The most reliable fertilizer test is made by field trials on the farm. Everyone using fertilizers should do a certain amount of experimenting. When fertilizers are applied as a general treatment, it is desirable to leave an untreated strip through the field. Similarly, an occasional strip through the field can be given a double application of fertilizer to make sure that the rate of treatment is adequate. If a deficiency of a single nutrient is believed probable, that nutrient can be applied as a strip through the field in addition to the regular fertilizer application. If two additional materials are being tried, it is advisable to make the strips so that they partially overlap. Thus the effect of each nutrient and the two nutrients in combination can be observed.

A careful record should be made of the exact location of the treated area. Growth can then be compared with that on adjacent areas. The *residual effect,* from portions of the fertilizer not used by the first crop, may influence crop growth during the next year.

Summary

Soils differ in their ability to supply nutrients to plants. When a soil is low in nutrients, it is usually profitable to supply them with fertilizers. Making a soil test is a convenient and relatively accurate method of determining which fertilizers to apply. Farmers can save hundreds of dollars by using the proper fertilizers.

The purpose of soil tests is to estimate the ability of soil to supply nutrients under expected conditions. There are different tests for phosphorus, potassium, nitrogen, and soil acidity and alkalinity. In most states, soil testing services are available. The farmer's job is to supply these with accurate samples of his soil. Samples should be made up of from 15 to 20 sub-samples taken from representative parts of a field. After the sub-samples have been thoroughly mixed and dried at room temperature, one pint should be sent to the laboratory. Farmers growing deep-rooted crops need to take subsoil samples, at a depth of 12 to 18 inches. A map should be made of the field to show the areas represented. The farmer should also supply the laboratory with a cropping history of the field. The laboratory will adjust fertilizer recommendations to offset acidity or alkalinity. Many factors, such as nutrient needs of planned crops, need to be taken into consideration when interpreting soil tests.

Soil testing kits are used by farmers and others to make spot checks on soils during the growing season. Plant tissue tests are a means of testing plants to find out what nutrients they contain. These tests are useful to check on the plant as an extracting agent, but results are difficult to apply except to the succeeding crops.

The most reliable fertilizer test is made by field trials on the farm. When fields are fertilized, one strip should be left untreated, another can be given a double application, a third might be given extra amounts of a single nutrient. Careful records must be kept.

Study Questions

1. How can soil testing be helpful to farmers?
2. How many soil tests are made each year in the United States?
3. What is the advantage of a soil test over a plant tissue test?
4. What important advantage do plant tissue tests have over soil tests?
5. What plant nutrients have received the most attention in soil tests?
6. What kinds of testing services are available to farmers? What is available in your area?
7. What are the values to be gained from soil testing kits?
8. How should you take a soil sample?
9. What factors are considered in making fertilizer recommendations based on soil tests?
10. Would you rely completely on soil and tissue tests to develop your fertilizer program, or use fertilizer field trials as well? Why?

Class Activities

1. Arrange for a trip through a nearby soil testing laboratory.
2. Make a field trip to practice taking and recording soil samples.
3. Demonstrate testing for soil nutrients and soil reaction, using a commercial kit.
4. Use a soil testing kit to test fertility of soil samples taken from various parts of the field, such as (1) fence rows, (2) dead furrows, and (3) fertilized rows.
5. Survey the kinds and amounts of fertilizers used in the community. Contact fertilizer dealers, county agents, and others.
6. Study the fertilizer-use experiences of farmers in the community.
7. Send fertilizer samples to official testing laboratories.
8. Demonstrate tissue tests on house plants or field crops grown in pots or boxes.

Commercial Fertilizers

Once the farmer has sent his soil samples to the laboratory for testing and has received a scientific analysis of the soils on his farm, he is ready to choose his commercial fertilizers.

You have already learned in Chapter 8, "Elements Essential to Plant Growth," that certain chemicals are required by growing crops. The major constituents of plants are carbon, hydrogen, and oxygen. These are provided by air and water. The other nutrients essential for plant growth come from minerals in the soil. The most important of these are nitrogen, phosphorus, and potassium. Minor, but still essential, elements include zinc, manganese, boron, copper, iron, molybdenum, calcium, sulfur, chlorine, and magnesium. In areas where the supply of one or more of these minor elements is deficient, their application to the soil or the plant is necessary. By using pure chemicals in nutrient solutions, scientists have demonstrated that plants cannot grow without each of the elements mentioned above.

Once it was discovered that plant growth could be increased by supplying the soil with elements in which it was deficient, the basis for a chemical fertilizer industry was established. The industry started to develop a little over one hundred years ago. It has now expanded so that the United States annually uses over 20 million tons of fertilizers.

Good farmers feed their soils as carefully as they feed their livestock. Crop yields have been doubled and tripled through the use of commercial fertilizers, in conjunction with other soil management practices. Production efficiency also increases.

Figure 13-1: Testing Fertilizer. Established in 1843, the Rothamsted Experimental Station near London, England, is one of the oldest soil management stations in the world. Note the many small, experimental plots where fertilizers and cultural practices are being tested. In the picture opening the chapter, you can see the famous Broadbalk Wheatfield at the station. Here, fertilization has maintained high yields for over a century. Fertilizer, equipment, and time are expensive. The farmer who carefully tests his soil to determine when and which fertilization is needed has a huge cost advantage.

What Are Commercial Fertilizers?

Many soils are deficient in more than one plant nutrient. Such deficiencies can be met by using two or more fertilizer materials. However, it is more usual to meet complex soil deficiencies by using mixed fertilizers. These are prepared by mixing various carriers of nitrogen, phosphorus, and potassium so that the nutrients are present in the desired proportions. They are sold on the basis of grade, and the minimum content of each nutrient is guaranteed to be present in the mixture. The grade is printed on the bag. An example is *10-20-10.* The first figure refers to the content of nitrogen; the second, phosphorus; and the third, potash. Actually, 100 pounds of 10-20-10 contains 10 pounds of nitrogen, 20 pounds of phosphoric acid (P_2O_5), and 10 pounds of potash (K_2O). Some mixed fertilizers have minor elements added to them. The guaranteed analysis will also show these. State laws regulate the sale and manufacture of fertilizers.

Minor elements should be added when needed, but they should not be used on more than an experimental basis unless a need is established. The minor elements are used in very small amounts. Use of even a few pounds per acre of some of these materials is toxic.

High-analysis materials are frequently the cheapest source of plant nutrients that the farmer can buy. The price per ton of high-analysis fertilizers is comparatively high, but not in terms of plant nutrients contained.

Pesticides in Fertilizers. An increasing practice is to add pesticides to mixed fertilizers. Pesticides can control weeds, fungi, and insects. The combined application saves time and expense.

Many fertilizer companies employ representatives who work with growers and dealers to discover and publicize the best methods of fertilizer application.

Amount of Fertilizer to Apply

Good farmers know that each farm—and in many cases, each field in each farm—presents its own individual fertility problem. This is why it is so important to collect accurate soil samples for laboratory analysis, and to check the results of the analysis with field trials. Fertilizer recommendations vary from area to area and from crop to crop. Specific recommendations must be secured from local sources. The best farmers test even these against results in their own fields with particular crops.

Fertilization for Maximum Yields. There have been many studies to determine the best rate of fertilizer application to a specific soil. There have been comparatively few attempts to adjust all growth factors except climate to secure maximum crop yields. Such studies necessitate having all nutrients at a satisfactory level and in proper balance and having physical factors such as aeration and moisture conditions near optimum. No material can be present in toxic concentrations. Disease and insects must be controlled.

These experiments, if successful, would determine the yield potential of a crop. While farmers would rarely try to achieve maximum yields, they, and research workers, would benefit by an idea of how closely to its potential a crop is yielding.

Figure 13-2: Maximum Yield? The 304.8 bushels of corn per acre produced on this field in 1955 is the highest corn yield recorded as of this writing.

Figure 13-3: Trees, Grass, and Field Crops. To achieve maximum profits, the operators of well-managed farms choose their fertilizers according to land use and crop needs.

Corn contests have been well documented, with yields above 200 bushels per acre recorded. Less well-documented studies show wheat yields of over 100 bushels per acre and oat yields of over 180 bushels per acre. These yields are much above the usual, but, again, it is not known whether they approach the maximum that might be obtained. To approach maximum yields for a crop, the stand of plants must be near optimum. As the fertility of a field is improved, the optimum stand of plants per acre also goes up. A high-fertility field should have more plants per acre than a low-fertility field.

Fertilization for Maximum Profits. In buying fertilizer, the farm operator is making an investment for profit. The yield increase must more than pay both the cost of the fertilizer and the cost of application. If a nutrient does not profitably increase yield or quality of a crop, it is not a good investment. Information from soil tests, past experience with fertilizers, and the fertility requirements of the crop should all be given consideration.

The amount of fertilizer that should be applied depends on the nature of the soil, the crop to be grown, the analysis of the fertilizer used, and the value of the crop. On some high-value crops grown on relatively infertile soil, the fertilizer application may exceed a ton per acre, whereas in areas of limited rainfall on soils with high native fertility but low potential yield, the best application rate may be only a few pounds per acre. This makes it impossible to give general recommendations on the amount to apply. Your state agricultural experiment station is the best source of reliable recommendations. This information will usually be available from your vocational agriculture teacher, county agent, or Soil Conservation Service personnel.

The greatest return on an investment is secured from the use of minor elements on deficient soils, because in some instances very light applications meet the needs of plants. In most cases where growth responses can easily be seen from the use of fertilizers, the increased crop growth will pay for the fertilizer. However, many factors must be given consideration in evaluating the return.

As the amount of any one plant nutrient is increased, the *return per unit* of application decreases. When a field is extremely deficient in any nutrient, the first unit of fertilizer will give a good yield increase. The next unit will provide a smaller increase, and the decline will continue until the yield increase will no longer pay the cost of the fertilizer. It is sometimes stated that two pounds of nitrogen per acre will produce an additional bushel of corn. This statement is not always true; it merely provides an indication of yield responses that might be obtained.

The fertilizer cost per-unit-of-yield-increase gets greater as we approach the maximum yield, until additional applications of fertilizer become unprofitable. Yield increases may continue even after the most profitable rate of fertilizer application is reached. If the fertilizer rate is a little too heavy, the additional growth from the last unit almost pays its cost. Similarly, if the rate is a little too light, the loss in return is not great.

When more than one nutrient is deficient, it becomes more difficult to obtain the optimum rate of each nutrient. Usually the yield increase from a unit of one nutrient is not the same as that from another. Thus, obtaining the optimum rate of each nutrient is difficult. In addition, the cost of the different nutrients is not the same. The lower the cost of fertilizer in relation to the value of crops, the more fertilizer can be applied profitably.

In spite of the foregoing qualifications, however, the effect of the use of fertilizers on farm profits is often greater than superficial examination would indicate. This is because many of the

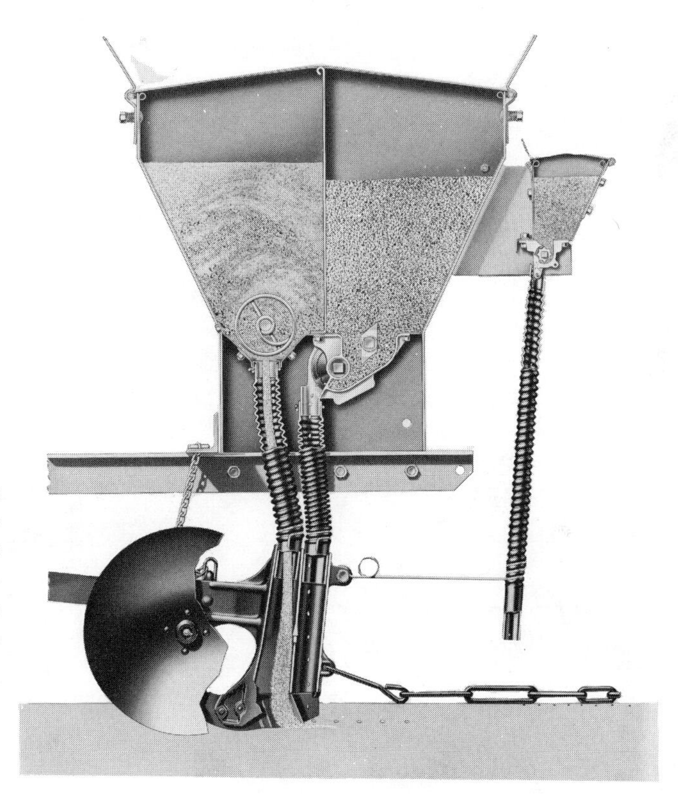

Figure 13-4: Fertilization and Seeding. This multipurpose drill, shown in cross section, places fertilizer where it will be immediately available to emerging roots.

costs of production are fixed. The costs of land, taxes, seed, and so on will be identical whether the crop yields are low or high. Thus a $20.00 increase in the value of a crop after a $10.00 investment in fertilizers may almost double the actual profit.

Effects of fertilizer on the maturity of the crop often affect the return from fertilizer use. An early crop may command a premium on the market. Also, the effect on quality may be an important factor in marketing.

Time of Fertilizer Application

Fertilizers are applied to stimulate plant growth. The nearer the application to the time the nutrient will be used by the crop, the better. Fertilizer nutrients can be lost by leaching or by reverting to unavailable forms. Under certain circumstances, for instance, nitrogen compounds can be reduced to elemental nitrogen. To get the maximum fertilizer efficiency, frequent small applications may be desirable even though frequent applications are costly. With many crops, there may be only one season when an application can be made.

For annual crops, the fertilizer application is frequently made at or near seeding time. In regions where leaching is not a problem, application may be made at a season when other work is light, such as late in the fall or during the winter.

Using Starter Solutions. The practice of using starter solutions gives "transplants" an early start. Transplants may include tomatoes, cabbages, peppers, and sweet potatoes. Starter solutions are

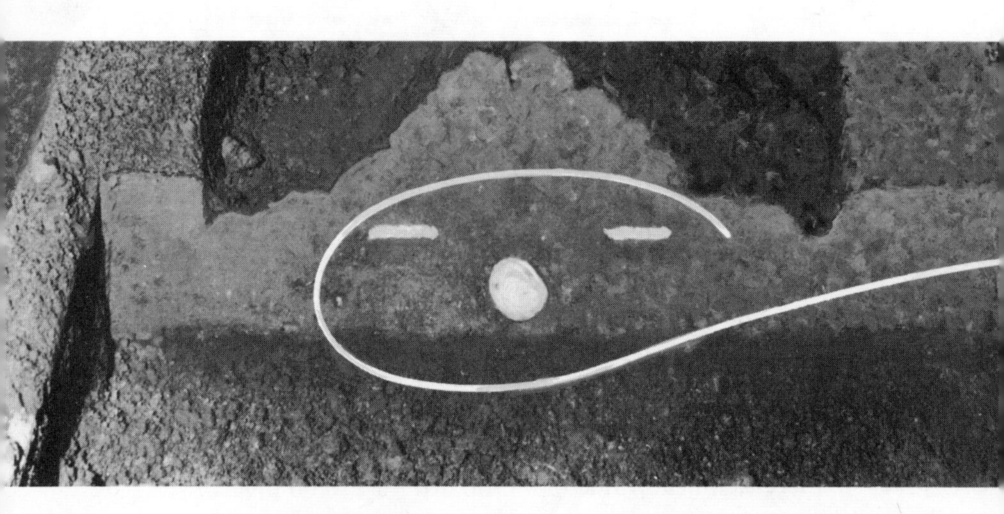

Figure 13-5: Banding. Separately illustrated are fertilizer bands above and below new-planted potatoes.

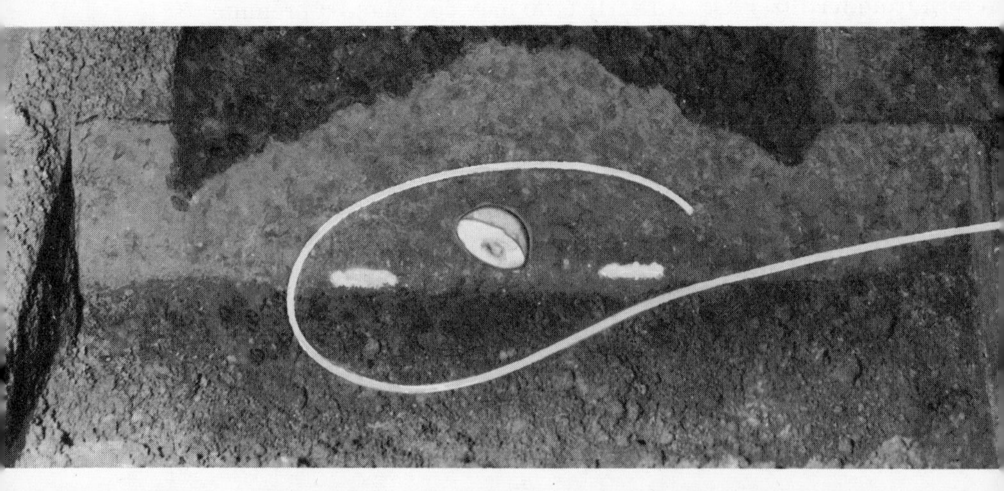

made by dissolving two to ten pounds of fertilizer, depending on analysis, in 100 gallons of water. Local recommendations should be followed on their preparation and use. Caution should be observed to prevent injury to roots. Starter solutions stimulate early maturity, because cold soils in the spring are apt to decrease available nitrogen.

How and Where to Apply Fertilizer

A few years ago, all commercial fertilizers were in the form of solids. At present, all of the major fertilizer nutrients also can be secured in liquid form. The use of liquids is increasing rapidly. Solids and liquids have essentially the same effect on the crop if applied in a comparable manner. Convenience usually determines which will be used.

Solid Fertilizers. Most solid fertilizers are granulated or pelleted for easy application and resistance to caking. Bags also are resistant to moisture. Commercial fertilizers should have careful storage and handling. If not kept dry, they are likely to cake.

There are many ways in which dry fertilizers can be applied. Some form of drilling is probably the most common method. This can be done either in conjunction with the seeding operation or separately.

Several problems face the farmer in determining where to place dry fertilizer in the soil. Fertilizers out of reach of the root systems of plants will not aid the growing crop. Growers may find it profitable to vary their fertilizer placement. Some crops benefit from more than one fertilizer placement. Common methods of dry fertilizer placement and application are: (1) drilling in with the seed, (2) banding, (3) placing fertilizer in the plow sole, (4) side-dressing, (5) broadcasting, and (6) spreading with manure. Detailed discussions follow.

Drilling In with the Seed. Under some conditions, the fertilizer is drilled into the soil with the seed. This practice is used with small grains. Fertilizer so applied may injure seedlings of some crops. A grower should experiment with field trials. Agricultural colleges make recommendations on rates of fertilizer application by this method in different sections of their state. Fifty to two hundred pounds per acre may be the upper limit.

Some growers will find it profitable to use a combination of fertilizer placements, such as drilling in with the seed and banding some distance from the seed. Drilling with the seed will help the plants to avoid damage from insects and soil-borne plant diseases.

Banding. By banding, the fertilizer is drilled in bands beside, above, or below the seed. It is effective with corn, potatoes, sugar beets, beans, and vegetables. When fertilizer bands are placed one or two inches to the side and below the seed, the close application makes fertilizer available to the plant's root system within two or three days after germination.

Crops with lateral roots benefit from fertilizer bands level with the seed rather than below it. Favorable results have been reported from placing such bands only on one side.

Placing Fertilizer in the Plow Sole. By placing fertilizer in the plow sole, fertilizer is placed in the soil at the time of plowing. Fertilizer hoppers are attached to the plow. The fertilizer is run through tubes to the bottom of the furrow. Thus the fertilizer is not mixed throughout the soil, but is placed deep. There it is available to the deeper-growing roots in zones of more favorable soil moisture. Fertilizers placed near the surface may be unavailable to crops under drought conditions.

Sidedressing. By sidedressing, fertilizer is fed into the soil on one or both sides of growing row crops. The technique is necessary because nitrogen fertilizers are subject to leaching; if applied all at one time, at planting, they may not be available when needed by the plant. Thus they are applied both at planting and at the second cultivation. Sidedressing is helpful in sandy soils, in soils low in organic matter, and in soils leached by heavy rains.

Broadcasting. A broadcast topdressing of fertilizer is used on meadows and pastures. Nitrogen fertilizers are most effective there, because nitrogen fertilizers penetrate the soil more easily than potassium or phosphorus.

Fertilizer broadcast on the surface and plowed under is mixed into the soil to plow depth. The technique is especially useful because nitrogen fertilizer hastens the decay of crop residues by favoring the growth of soil bacteria.

Fertilizers broadcast on the surface and harrowed in are mixed only with the surface soil. The method is used with small grain and grass crops. Phosphorus fixation is higher as a result of more thorough mixing. Harrowing makes poorer use of fertilizer than such methods as banding or mixing with the seed.

Spreading with Manure. Fertilizer also can be applied by spreading it uniformly over the top of a spreaderload of manure. From the number of spreaderloads that are applied per acre, it is easy to calculate the amount of fertilizer applied.

Liquid Fertilizers. Liquid fertilizers are coming into more common use. The liquids are frequently sprayed on the surface with standard spray equipment. They also are dribbled on the surface by a gravity feed on a boom fed from a tank. To get uniform distribution using this method, it is necessary to compensate for pressure changes as the amount of the solution in the tank decreases. Liquid fertilizers can be distributed with irrigation water. The fertilizer is usually injected into the water a little ahead of the location where water-spreading starts, so that the flowing water brings about thorough mixing. Distribution is only as uniform as water application.

Some of the liquid fertilizers have ammonia gas in them, and this is volatile (turns easily into vapor). Such materials need to be injected into the soil to a depth of two inches by passing the solution into the soil through a "shoe," much like that used on a seed drill. However, when ammonia solutions are injected into irrigation water, the loss from gases is very small.

Fertilizer solutions are corrosive, and resistant materials must be used in handling them. The manufacturers' recommendations

Figure 13-6: Injecting Anhydrous Ammonia. Although anhydrous ammonia is stored as a pressurized liquid, it reverts to a gas when pressure is released and thus must be injected into the soil to prevent its escape. Here, injection is being combined with cultivation of trashy summer fallow with a rod weeder.

with regard to protection against corrosion are usually satisfactory and should be followed.

Most nitrogen fertilizers are soluble enough to be prepared as liquids. High-analysis materials, such as ammonium nitrate and urea, are usually used in liquid fertilizers. The potassium fertilizers also can readily be applied as a solution. The only phosphate fertilizer compounds soluble enough to be used satisfactorily in liquid form are phosphoric acid and ammonium phosphate. Frequently some ammonia gas is added to liquid fertilizers, but precautions are needed in application and handling to prevent its loss.

Since the solubility of most salts decreases with decreases in temperature, there is a tendency for some of the material in liquid fertilizers to "salt out" at low temperatures if the solution is too concentrated. When salting-out occurs, the concentration of fertilizer in the solution is decreased. The solids in the solution also interfere with application.

Fertilizer solutions contain water and are consequently lower in analysis than the solids from which they are made. Because of this, the solutions are usually made up locally rather than shipped long distances. Solutions can readily be handled with pumps and gravity flow. This convenience of handling has increased their use.

Kinds of Commercial Fertilizer

The following discussion deals with the kinds of nitrogen, phosphorus, and potassium fertilizers produced by the industry.

Nitrogen Fertilizers. Nitrogen in its natural state is a colorless, gaseous, tasteless, and odorless element which makes up about four fifths of the earth's atmosphere by volume. There are about 34,500 tons of nitrogen over each acre of the earth's surface. The element is an essential constituent of living plant and animal cells From the air, it is made available to plants by nitrifying soil bacteria. About 90 per cent of commercial nitrogen is also manufactured from the air. Major nitrogen fertilizers are as follows:

Sulfate of Ammonia ($(NH_4)_2SO_4$). Commercial grades of ammonium sulfate contain 20 to 21 per cent nitrogen and about 24 per cent sulfur. Ammonium sulfate does not readily absorb water and this tends to keep the fertilizer in good condition.

Figure 13-7: Maintained Without Nitrogen Fertilizers. This fine stand of alfalfa has been maintained for five years by means of an annual application of 500 pounds of 0-15-30. Farmers save money by knowing what nutrients their crops and soils need.

Since it leaves an acid residue in the soil, it is suitable to use under alkaline conditions or where acidity is desired. Its use on acid soils makes them more acid, but this can be corrected by adding lime.

Nitrate of Soda or Sodium Nitrate ($NaNO_3$). Nitrate of soda is one of the oldest nitrogen fertilizers. Its biggest source has been the natural deposits in Chile, but it is also made synthetically. Nitrate of soda is an ideal nitrogen fertilizer since most plants can absorb nitrates directly. This fertilizer leaves an alkaline residue, which may be an advantage in acid soils. Nitrate of soda needs to be stored in dry places, because it readily absorbs moisture. It contains about 16 per cent nitrogen.

Ammonium Nitrate (NH_4NO_3). Ammonium nitrate is most available to plants. Some manufacturers combine it with dolomitic limestone. The combination furnishes calcium and magnesium as well as nitrogen.

Safety precautions must be followed in handling and storing nitrate fertilizer. Suggested rules are:

1. Do not smoke or use open flame in or near space where ammonium nitrate is stored.
2. Keep ammonium nitrate away from steam pipes, electric wiring, and combustible materials of all kinds.
3. Store in a well-ventilated building. Large quantities should be stored several hundred feet from other farm buildings.
4. Clean up spilled ammonium nitrate. Do not return contaminated material to the bag.
5. Promptly destroy empty bags that have contained ammonium nitrate. Such bags are highly inflammable.

Ammonium Phosphate ($NH_4H_2PO_4$). Pure ammonium phosphate contains 61.71 per cent phosphoric acid and 12.1 per cent nitrogen. This fertilizer has proved excellent for topdressing grasses and legumes. Its use increases soil acidity, and on acid soils it may need to be applied with lime.

Calcium Nitrate ($Ca(NO_3)_2$). Calcium nitrate carries about 17 per cent nitrogen and 34 per cent lime. It helps decrease soil acidity.

Urea-Form Synthetic Nitrogen. Urea-form synthetic nitrogen is water-insoluble. In the soil it changes to ammonia and then to nitrate. It is long-lasting, like barnyard manure. Urea-form is suitable for use on lawns, golf courses, and grass crops.

Anhydrous Ammonia. Anhydrous ammonia is a colorless, pungent gas at usual temperatures and pressures. However, under high pressure it changes to a liquid and can be stored in this form. Containing 82 per cent nitrogen, it is probably the most concentrated nitrogen fertilizer developed. It is usually purchased as a liquid in heavy metal tanks which will withstand the necessary pressure. When the pressure is released with a valve and the liquid changes back to a gas, it must be injected into the soil to a depth of about six inches to allow it to be absorbed without escaping to the air. Anhydrous ammonia can also be dissolved in irrigation water. It should be applied to friable soils. The soil should be neither too dry nor too wet.

Aqua Ammonia. The nitrogen content of aqua ammonia is about 40 per cent. Aqua ammonia solutions require low vapor pressures—usually 25 pounds per square inch at 104° F. The low storage pressure of aqua ammonia makes it comparatively safe for handling and application to the soil.

Phosphate Fertilizers. American farmers use more phosphate than any other fertilizer. The most common source of phosphate fertilizer is rock phosphate. The United States has the largest known deposits. They are in Montana, Idaho, Utah, Wyoming, Florida, and Tennessee.

Pure phosphorus is a greyish metallic substance that bursts into flame and burns with intense heat when exposed to air. Phosphorus in fertilizer is combined with other elements to make it usable. It hastens plant growth and the development of fruits and seeds. The phosphorus content of average soils is about one-half that of nitrogen and one-twentieth that of potassium.

Figure 13-8: Phosphates for Quick Growth. Phosphates are especially valuable in promoting growth of young plants. Lands requiring intensive conservation practices, such as the contour strip cropping illustrated, benefit from the earlier ground cover provided by phosphate fertilizers.

Most of the phosphate fertilizer that is applied to soil is converted to insoluble compounds of iron and aluminum phosphate. Clays and clay loams do this more readily than sand and sandy loams. In many cases, not more than 10 per cent of the phosphorus applied to soils is available to crops the first year. However, young plants are greatly benefited, and phosphorus not absorbed in the year of application is gradually released. Responses four to five years after application are not uncommon.

Superphosphate. Superphosphate is the most widely used phosphate fertilizer. Superphosphate is made by treating ground rock phosphate with sulfuric acid. Superphosphate fertilizer contains 18 to 22 per cent P_2O_5, 20 per cent calcium, and 12 per cent sulfur.

Triple superphosphate is a product that varies from superphosphate in that it contains 40 to 45 per cent available phosphoric acid, 12 per cent calcium, and 1 per cent sulfur.

Rock Phosphate. Considerable ground rock phosphate is used directly as a fertilizer on acid soils. The finer the material, the greater the availability of its phosphorus. On neutral and alkaline soils, however, it is not satisfactory.

Basic Slag. Basic slag is a finely ground by-product of iron smelting by the basic Bessemer and open-hearth processes. Basic slag contains lime in addition to phosphorus. Like rock phosphate, it is useful on acid soils, especially muck and grass sod.

Ammonium Phosphate. Ammonium phosphate is a source of both nitrogen and phosphorus. The nitrogen content usually varies between one fourth and one half the P_2O_5 content. It is one of the most readily available sources of phosphate.

Calcium Metaphosphate. Calcium metaphosphate—$Ca(PO_3)_2$—is a comparatively new product on the market. It contains over 60 per cent P_2O_5. On acid soils it is about equivalent to superphosphate, but on alkaline soils its phosphorus is not readily available to plants.

Phosphoric Acid. Phosphoric acid is a solution, and its use is increasing rapidly. It is a strong acid and must be handled with care. It is a highly available source of phosphorus.

Potash Fertilizers. Many soils are lacking in available potash, even though the average plow layer contains 40,000 pounds per acre as compared to 1500 pounds of phosphorus. Pure potassium is a whitish-gray metal that bursts into flame when exposed to air. Potassium in fertilizer is combined with other elements. Potash salts are mined in New Mexico, and are extracted from alkali lake waters in California and Utah.

Muriate of Potash or Potassium Chloride (KCl). Muriate of potash usually contains 60 to 62 per cent soluble potash. This is our most important potash carrier in the United States. However, the chlorine content of potassium chloride is thought to be injurious to some crops, such as tobacco.

Sulfate of Potash or Potassium Sulfate (K_2SO_4). Average grades of sulfate of potash contain 90 per cent potassium sulfate, which is equivalent to 48.7 per cent potash.

Nitrate of Potash or Potassium Nitrate (KNO_3). Nitrate of potash is largely a by-product of refined nitrate of soda deposits in Chile.

Potassium Magnesium Sulfate. Potassium magnesium sulfate contains the equivalent of 21.75 per cent potash and 18.75 per cent magnesium oxide. This fertilizer is a highly available source of both potash and magnesium.

Nitrate of Soda and Potash, or Potassium Sodium Nitrate. Nitrate of soda and potash contains not less than 14 per cent nitrogen and 14 per cent potash.

Summary

Once the farmer has received a scientific analysis of the soils on his farm, he is ready to choose his commercial fertilizers. Major elements supplied by fertilizers are nitrogen, phosphorus, and potassium. Minor elements that are also essential for plant growth and that may need to be added to the soil are zinc, manganese, boron, copper, iron, molybdenum, calcium, sulfur, chlorine, and magnesium. Fertilizers are graded according to the nitrogen, phosphorus, and potash they contain; a 10-20-10 fertilizer contains 10 pounds of nitrogen, 20 pounds of phosphoric acid, and 10 pounds of potash for each 100 pounds of gross weight. Minor elements are also indicated on the bag. Some fertilizers contain pesticides.

It is not known how much fertilizer would have to be applied to obtain the best possible yield of any particular crop. The farmer's consideration in buying fertilizer is not maximum yield, but maximum profit; there is a point beyond which fertilization will not improve crop yields enough to pay the cost of the extra fertilization. In general, however, the use of fertilizer greatly increases farm profits.

igure 13-9: Potash De-iciency. Five years ear-er, this field was seeded vith Ladino and orchard rass. Only orchard rass now remains, be-ause of a potash defi-iency in the soil.

Fertilizer should be applied at the time when it will be most valuable to growing crops. For many crops, fertilization is done at seeding time. Starter solutions are used with transplants.

Common methods of applying dry or solid fertilizers are: (1) drilling in with the seed, (2) banding, (3) placing fertilizer in the plow sole, (4) sidedressing, (5) broadcasting, and (6) spreading with manure. Liquid fertilizers may be sprayed on the surface with standard spray equipment, dribbled on the surface by a gravity feed on a boom, injected into irrigation water, or injected into the soil.

Many kinds of commercial fertilizers are available to farmers.

Study Questions

1. How much commercial fertilizer is used nationally each year?
2. What is the meaning of the fertilizer grade 10-10-20?
3. What factors would have to be controlled to determine maximum yields?
4. What economic factors should a farmer consider in buying fertilizer?
5. At what time of the year should fertilizer be applied?
6. What are the methods of applying *solid* and *liquid* fertilizers?
7. What is the purpose of the different placements of fertilizer in the soil, and what relation do these placements have to the seed and the growing plants?
8. What kinds of nitrogen fertilizers are available?
9. What kinds of phosphorus fertilizers are available?
10. What kinds of potassium fertilizers are available?

Class Activities

1. Secure exhibits of pure phosphorus and potassium from the school chemistry department and arrange a demonstration of their properties.
2. Make a study of the amounts and kinds of commercial fertilizer used locally.
3. Determine the need for use of commercial fertilizers on local farms.
4. Arrange for demonstrations of different methods of fertilizer application.

Farm Manures

Another important means of enriching soil is the use of barnyard manure. It has been estimated, on the basis of the increase in crop yields realized on fields to which manure has been applied, that manure is worth from three to four dollars a ton.

On the average, each ton of manure contains ten pounds of nitrogen (N), five pounds of phosphate (P_2O_5), and ten pounds of potash (K_2O), as well as small quantities of other essential nutrients. Except when livestock are raised largely with feed bought off the farm, manure alone cannot maintain soil fertility. So many nutrients are removed from the soil by the harvest of crops and livestock that the amount returned by applications of manure usually must be supplemented with lime and commercial fertilizer. The value of manure as a source of nutrient elements lies in the fact that it permits the farmer to receive double value from a considerable portion of his soil nutrients—once when he first grows crops with them, and again when he returns them to the soil in the form of manure.

Of even greater importance to the farmer is the value of manure as a source of organic matter and microorganisms. Each ton of manure not only contains partially decomposed plant materials and dead bacteria, but millions of live bacteria ready to convert these into humus and available plant nutrients. The live bacteria in manure also help convert soil minerals and chemical fertilizers into forms available to plants. Manure improves soil tilth, structure, and moisture-retaining capacity. Barnyard manure is the most important source of organic matter on many farms.

Manure applied as a topdressing gives the further benefit of retarding wind and water erosion.

In all of these ways, then, manure helps the farmer. Good farmers know that the wise care and use of manure often turn a potential loss into a fine cash profit.

Properties of Different Manures

Different farm animals produce manures of different fertilizer content. Poultry manure has up to three times the nitrogen content of other manures. Sheep manure is twice as rich as cattle manure in nitrogen and potash, and swine manure is twice as rich as cattle manure in phosphate.

Manures from horses and sheep ferment and heat more readily than those from hogs and cattle. Farmers frequently refer to cattle and hog manures as "cold" manures for this reason.

Figure 14-1: Cattle and Litter. This fine young stock is kept clean and healthy with a wood chip litter which conserves liquid manure. Which nutrient elements are concentrated in liquid manure?

In all manures, most of the phosphate is in the solid excrement, while most of the potash and nitrogen are in the urine. A pound of urine contains about five times as much potash and from two to three times as much nitrogen as the solid portion of manure. Poultry manure is the most concentrated because the urine is voided in a mixture with the solid excrement.

The richness of manure also varies with the amount and kind of bedding used. The kind and quality of feed given to livestock is another factor. Manure from livestock feed lots will be high in nutrients because the feed is high in nutrients.

Conservation of Manure

As we have seen, manure is an important asset to the farm. It can help maintain and improve the fertility and physical condition of the soil. Manure needs to be properly conserved if its great potential value is to be realized. Two key practices are (1) the conservation of liquid manure, and (2) the prevention of rotting before manure reaches the fields. Rotting results in loss of nitrogen in manure. A valuable improvement project on many farms would be the practice of manure conservation. Up to half of the potential value of manure is often lost. The possible dollar gains from proper handling are tremendous.

Conservation of Liquid Manure. The liquid part of manure may be conserved by tight gutters, by floors on feed lots, and by litter of straw, peat moss, sawdust, shavings, or wood chips. Litter also keeps animals dry and comfortable.

Straw Litter. Grain straw is the most common form of livestock bedding. Recommended requirements of straw litter per animal per day are ten to twelve pounds for horses, nine pounds for cattle, 1½ pounds for hogs, and one pound for sheep. Unchopped straw is preferable to chopped straw because it stays in place better. Other types of litter are sometimes used for dairy cattle because straw litter has been found to contain bacteria that can contaminate milk.

Peat Moss Litter. Peat moss is a good litter because it absorbs liquids in large amounts and because it is high in acids, which discourage harmful bacteria. In poultry houses, however, peat moss may cause considerable dustiness.

Figure 14-2: Woodchipping Machine. Wood chips prepared on machines like this are used for mulches as well as litters.

Sawdust, Shaving, and Wood Chip Litter. Sawdust, shavings, and wood chips compare favorably with other forms of litter in their capacity to absorb liquids and thus to save the nitrogen and potash in manure. Wood chip litter is prepared by woodchipping machines which can take logs up to six inches in diameter. The knives on these machines can be adjusted to produce chips in sizes suitable for barnyard or poultry-house use. Sawdust, shavings, and wood chips are sometimes also used as mulches.

Prevention of Rotting. Under some conditions, rotting of manure may be desirable, because well-rotted manure can be worked easily into the soil. Ordinarily, however, rotting should be prevented. Rotted manure has lost much of its potential nutrient value. It often has the added disadvantage of containing large amounts of dry straw, which can separate soil layers and thus cause soil dryness.

When manure rots or ferments, much of its nitrogen is lost in the form of gas. Strong ammonia odors in barns, feed lots, and chicken coops indicate such losses.

Since the rotting of manure is a process of decay, rotting can be prevented if the manure is kept moist. Tight gutters, feed lot floors, and litter, discussed above, will help prevent rotting.

The use of *nitrogen stabilizers* is another effective means of conserving manure. Many farmers spread superphosphate (0-20-0) on manure, litter, and gutters. Superphosphate is high in gypsum (calcium sulfate). The gypsum changes the ammonia of the manure to ammonium sulfate, which is soluble but less volatile and more stable than ammonia. Gypsum can also be used directly, in place of superphosphate, for this purpose. Triple superphosphate (0-45-0) and calcium metaphosphate (0-63-0) contain little if any gypsum and thus are less desirable. Since manure contains less phosphate than nitrogen and potash, however, all phosphate materials tend to make manure a more balanced fertilizer.

Ground limestone and hydrated lime may or may not function as stabilizers. Ground limestone ($CaCO_3$) spread on fresh manure neither reduces nor speeds rotting. On rotted manure it speeds loss of ammonia. Small amounts of hydrated lime on fresh or fermented manure greatly increase ammonia losses, but large amounts of the same material temporarily slow ammonia losses.

Accumulation and Storage of Manure

The best way of accumulating and storing manure is the use of *built-up litters.*

Livestock farmers use this method to let manure accumulate in pen-type barns or sheds or in covered lots. Built-up litters are especially helpful if the right amount of bedding is used. The straw will absorb the liquids. The packing down of the manure by the animals will retard fermentation losses. It should be remembered, however, that the manue will dry out soon after livestock are removed. This drying will increase ammonia losses. To prevent these, the manure should be spread on the fields as soon as possible, and plowed under or disked into the soil.

Built-up litters are also effective with poultry manure if superphosphate, gypsum, or (less desirably) limestone is spread at regular intervals. Superphosphate is the best agent. About 25 pounds should be used for each 100 square feet of floor space or for each 100 birds. The first application should be made a few days after the litter is started, with new applications for each few inches of additional litter.

Other Methods of Storage. When the use of built-up litters is not possible, manure should be stored in such a way that it will be kept compact and moist. A closed shed with watertight walls and floor is most effective. A tight-walled silo-like trench also gives good results. If it is necessary to store manure in piles, these should have flat tops and straight sides.

Application of Manure

Prolonged storage of manure is usually undesirable. The way to get the most good from barnyard manure is to load it from the barns onto a manure spreader and haul it directly to the field. A common practice is to add 40 to 50 pounds of 20 per cent superphosphate on top of each load. As noted above, this makes the manure a more balanced fertilizer. Other farmers add various proportions of a complete chemical fertilizer, thus making chem-

Figure 14-3: Paved Barnyard. Barnyard pavement helps prevent loss of liquid manure and makes collection easy.

ical fertilizer and manuring a single, more economical operation.

It is a good practice to spread manure during rainy periods, because the rain works the more soluble materials into the soil. During dry periods manure should be disked in or plowed under to prevent ammonia losses.

Rate to Apply. "Frequent" light applications of manure are applications every two or three years. In general, frequent light applications bring greater total returns than less frequent heavy applications. Applications of four to eight tons per acre are recommended. Poultry manure should be applied at the rate of three tons or less per acre because of its richness. Light applications of manure supplemented with fertilizer drilled in the rows near the crop will provide as good returns as heavier applications of manure alone.

In rotations, applications once every four or five years are desirable.

Other factors affecting application rate include the amount of manure available, the kind of soil, and the requirements of the crop.

Figure 14-4: Loading Manure. Modern equipment simplifies collection and loading.

Figure 14-5: Spreading Manure. Manure feeds soil microbes, adds organic matter, improves tilth, increases farm profits.

Where to Apply. Some soils and crops profit more from manure than others.

Soils Most Benefited. The poorest soils on the farm benefit most from manure. Sandy and severely eroded soils benefit more from manure than from commercial fertilizer, for instance. Being deficient in organic matter, such soils tend to lose commercial fertilizer through leaching. Manure will add organic matter. If supplemented with commercial fertilizer, manure also will help retain most of this fertilizer for plant use.

Alkali spots are another example of problem soils that benefit most from manure.

Crops Most Benefited. While all crops benefit from manure, row crops like potatoes, sugar beets, corn, and garden and truck crops respond especially well. In crop rotations, manure thus should be applied to the row crops. Crops just following manure application benefit more than crops that follow later in a rotation.

Pastures and meadows respond well to light topdressings. Because cattle may hesitate to graze manured areas, it is a good practice to cover one portion of the pasture at a time.

Topdressings of two to four tons per acre on fields planted to legumes help legumes get established.

Less beneficial are heavy applications of manure on land

planted to grain. These may stimulate heavy straw growth and cause lodging.

Summary

Barnyard manure is an important means of enriching soil. Manure is estimated to be worth three to four dollars a ton in increased crop yields. On the average, each ton contains ten pounds of nitrogen, five pounds of phosphate, and ten pounds of potash. While manure alone cannot maintain soil fertility, it provides the farmer with a means of receiving double value from a considerable portion of his soil nutrients.

Manure is even more important as a source of organic matter and live bacteria.

Different farm animals produce manures of different fertilizer content. Most of the phosphate is in the solid excrement, most of the potash and nitrogen in the urine.

Manure must be properly conserved if its great potential value is to be realized. The liquid part of manure may be conserved by tight gutters, by floors on feed lots, and by litter of straw, peat moss, sawdust, shavings, or wood chips. Ordinarily, rotting of manure should be prevented, because when manure rots it loses much of its nitrogen in the form of ammonia gas. Rotting may be prevented by liquid manure conservation and by the use of nitrogen stabilizers. Superphosphate and gypsum are favored materials for the latter purpose. The best way to accumulate and store manure is the use of built-up litters, which keep manure compact and moist. Manure may also be stored in closed sheds, tight-walled trenches, and piles with flat tops and straight sides.

Since prolonged storage is undesirable, manure is best hauled directly to the field. Many farmers make manure a balanced fertilizer by adding superphosphate or various proportions of complete chemical fertilizers. Manure spread during rainy periods tends to work into the soil. During dry periods, it should be disked in or plowed under. In general, frequent light applications are preferred; these are applications of from four to eight tons per acre every two or three years. In rotations, applications every four or five years are desirable. The poorest soils on the farm benefit most from manure. Row crops respond especially well. Manure is also recommended for pastures and legume crops, but heavy applications for grain are sometimes harmful.

Study Questions

1. What determines the dollar value of manure?
2. What are desirable kinds of manure litter?
3. What is the average fertilizer content of manure?
4. What are the approved practices in preventing ammonia losses in manure?
5. How do built-up litters conserve manure?
6. What is the value of hauling manure direct to the field?
7. What soils and crops are most benefited by manure?
8. What rates and times of application are most profitable?

Class Activities

1. Make a field trip to several farms to observe practices in conserving barnyard manure. Before the trip, prepare a list of things to observe or questions to be asked.
2. Consider plans of individual class members for better care of the barnyard manure on the home farm.
3. Figure the fertilizer value of a ton of manure using local fertilizer prices and estimating that a ton of manure includes 10 pounds N, 5 pounds P_2O_5, and 10 pounds K_2O.
4. Have class members check home use of approved practices in handling manure:
   _____ a. Conserve valuable liquid parts of manure by use of proper bedding, tight gutters, and floors on feed lots.
   _____ b. Lessen fermentation or rotting by use of preservatives such as superphosphate or gypsum.
   _____ c. Haul manure directly to the field when practical to prevent losses of ammonia.
   _____ d. Keep manure well packed with frequent applications of straw in feed lots and in storage piles.
   _____ e. Work manure into the soil soon after application.
   _____ f. Add superphosphate to the manure to make it a balanced fertilizer.
   _____ g. Make frequent application of four to eight tons per acre.
   _____ h. Use recommended soil and crop practices along with manure.

Land Drainage Needs and Practices

All growing plants require large amounts of water. Yet when the soil is so wet that even the large soil pores are flooded, most plants cannot grow satisfactorily. The problem is not too much water, but too little air in contact with roots. Like soil water, soil air is an essential plant nutrient, and plants will die if long without it. Soil microorganisms also require air for normal reproduction and growth, and suffer from excessive wetness. Even if soil is not wet long enough to kill a crop, heavy damage can result. For instance, soil that remains wet for prolonged periods in the spring encourages shallow root growth. If the soil later dries, as during a summer drought, the shallow-rooted plants will wilt.

In addition, wet soil generally has poor tilth. Wetness delays planting, plowing, and other cultural practices. It often increases tractor fuel consumption, and can strain valuable machinery.

For all of these reasons, when natural drainage is deficient and soil pores cannot rid the soil of free water, drainage must be provided by the farmer. In areas of high precipitation, such as the Eastern states, level lands often need some form of drainage. In the West, lands leveled for flood irrigation also need some provision for the removal of excess water.

Forms of Drainage

Drainage can take two forms: (1) surface drainage, and (2) subsurface or underground drainage.

Surface Drainage. In surface drainage, ditches are provided so that excess water will run off before it enters the soil. This is a comparatively simple method of correcting a drainage problem. However, the water intake rate of the soil should be kept as high as possible, so that water that could be stored will not be drained off. Otherwise, when rains are not frequent or properly spaced, the soil will become too dry.

Figure 15-1: Surface Drainage. Before ditching, this pasture was not suitable for any farming purpose.

Subsurface Drainage. Subsurface or underground drainage is much more difficult.

In naturally well-drained soil, subsurface drainage takes place because excess water passes through the soil to the water table, where it flows underground to the open streams of the area and is carried away. When there is a natural block to the flow of an underground stream, the water backs up and forms an underground lake. Eventually, this underground lake may reach the surface and form a marsh or swamp. Even if this does not happen, the underground lake may become near enough to the surface of the ground to interfere with plant growth. Then it becomes necessary for the farmer to provide underground drainage. Mole drains or tile drains are generally used for this purpose.

Charting the Water Table. In areas needing subsurface drainage, the water table is always close to the surface. Its height can be determined fairly easily. This is done by digging a series of shallow wells and recording the level at which water stands in them. These levels must be recorded at a "true" level and not simply at the depth at which water is present below the ground surface. A surveyor's level or some similar device is necessary for this.

Figure 15-2: Irrigation and the Water Table. Irrigation, here being accomplished with siphons and the unique canvas dam in the foreground, can raise the water table to a level that deprives roots of air and can injure them with salt accumulation. Subsurface drainage is a common corrective measure.

Figure 15-3: Using the Level. Use of the level, here being learned by boys at the Montana 4-H Conservation Camp, is an important aid in designing both drainage and irrigation systems.

The maximum heights of water table levels will occur when water is flowing into an area rapidly, as during a rainy season or in the spring. To allow for this fact, the water table levels should be taken over a considerable period of time and particularly during that period when water inflow is greatest. A study of *how* water is coming into an area also is an essential step in designing an efficient drainage system.

Designing a Drainage System

The objective in designing a drainage system is to get the simplest and cheapest series of drains that will allow crops to make satisfactory growth. The proper placement of drains is much more important than their number.

Small Systems. Where only a single small area needs to be drained, such as a wet spot which interferes with early seeding or cultivation of a field, a suitable plan can be determined by a field inspection. An adequate fall for gravity flow of water must be available. Some system needs to be used to establish a uniform grade or fall. This might be done with a hand level, a carpenter's level, or regular surveying instruments.

Large Systems. When planning a large drainage system, farmers need to consider the entire watershed or drainage area. An adequate drainage system often involves several farms. In such cases, a cooperative project is advisable. Where very large areas of bottomland have been drained, farmers have organized drainage districts.

For help in planning any extensive drainage system, farmers can avail themselves of the assistance of the Soil Conservation Service if they are members of a SCS District. Under any condition, a map should be drawn to show:

1. Sources of water to be drained off.
2. Boundaries and slopes of areas to be drained.
3. Existing drains or ditches.
4. Location and elevation of all swales or marshy areas, watercourses or natural drainage courses, knolls, and ridges.
5. Location and elevation of possible outlets.
6. Areas that will drain into each part of the system.

After the drainage system is laid out on a map, grades and sizes of drains need to be determined and costs estimated.

In deciding whether surface or subsurface drainage is to be used, the farmer should consider a number of factors. The installation cost of ditches is less than that of tile, but ditches are effective mainly for draining surface water, not for draining soil water unless deep, relatively expensive ditches are used. Ditches are the most practical means of draining tight soils which do not absorb water readily. Ditches are also needed where excess water accumulates even after underground systems have been installed. Tile drainage is cheaper in the long run when land is expensive, since subsurface drainage leaves the topsoil free for cropping.

Tile drainage leaves the soil in better condition for plant growth, except in the case of tight soil. Tile drainage requires less maintenance, needing little attention except to drains, while ditches must be periodically cleared of silt, weeds, and brush. Tile drainage is more convenient for farming operations, since there are no open ditches for machinery to cross. Farmers often use a combination of both systems, according to the needs of the particular farm.

Once these matters have been considered, the drainage system should be staked out in the field. Levels should be run for each of the proposed drains to determine accurately its grade or fall. Where natural ditch outlets are not available, a grade also must be determined for ditches to carry away the water absorbed by tile drains.

Figure 15-4: Drainage Ditch. More than 500 acres of profitable flatland are being added to this farm by installation of a surface drainage system.

Surface Drainage

Surface drainage is provided through open ditches. Field ditches empty into collecting ditches built to follow a natural watercourse. A natural grade or fall is needed to carry the water away from the areas to be drained. The location of areas needing surface drainage can be determined easily by observing where water is standing on the ground after a heavy rain. As stated above, ditches are not effective devices for removing *soil* water, unless very deep ditches are used. Ditches ordinarily will remove surface water only.

Design of Ditch Systems. Field ditches and collection or outlet ditches should be large enough to remove at least two inches of water in 24 hours from level to gently sloping land. The capacity of the drainage system should be based on the amount and frequency of heavy rains.

How quickly water runs into ditches depends on the rate of rainfall, land slope, the condition of the soil, and plant cover. The area that a ditch will satisfactorily drain depends on how quickly water runs into the ditch, the size of the ditch, its grade or slope, and its irregularity. The irregularity of a ditch is measured by its roughness and its content of debris and growing vegetation.

In level areas, a collecting ditch may be installed along one side and shallow V-shaped field ditches constructed to discharge into this collecting ditch. Field ditches used to discharge water into collecting ditches should be laid out parallel to each other, 50 to 150 feet apart. They should be 12 to 24 inches deep, depending upon the depth of the collecting ditch.

Angle of Ditch Sides. Field ditches are either narrow with nearly vertical sides or V-shaped with slightly sloping sides. V-shaped ditches are easier to cross with machinery. Large V-shaped ditches can be planted to grass, which can be mowed for hay.

V-shaped field ditches to be crossed by machinery should have side slopes of three-to-one (horizontal to vertical), or should be even flatter. Narrow field ditches may have sides as nearly vertical as will resist frost action and erosion by water running down the sides.

Figure 15-5: Drop. This check gate, which is part of an irrigation system, shares most of the features of the drop spillways used to prevent erosion in drainage systems. Note the headwall, concrete sidewalls, apron, and cutoff wall. Why have stones been put on the lower ditch bank?

For collection or outlet ditches in clay and clay loam soils, two-to-one side slopes are best. Vertical sides can be used in rock. In hardpan and loose rock, one-half-to-one slopes are satisfactory. In peat and muck, nearly vertical ditch-banks have stood up better in many places than flatter slopes.

Preventing Erosion. Sharp curves in ditches should be avoided to lessen erosion of banks. Soil erosion also may be a problem in ditches with too steep a grade or slope. A good practice may be to construct ditches in field sections. These relatively short ditches should have less fall than the natural slope, and should be connected with the ditches of the next field section by drop spillways. Drop spillways are recommended wherever the average velocity of flow will exceed 2½ feet per second in sand or sandy loam, three feet per second in silt loam, 3½ feet per second in sandy clay loam, four feet per second in clay loam, and five feet per second in stiff clay or fine gravel.

A drop spillway is needed at the bank of a large ditch wherever there is a considerable drop between two ditches. The headwall may be constructed of reinforced concrete, brick, or stone. To prevent cutting, the structure should have an "apron" or floor for the water to fall on. The apron should have side walls, and a cut-off wall at the outer edge.

Flumes. A grassed flume is a broad channel lined with grass or other vegetation. It is used to provide an economical means of safely admitting water from a field ditch into a collection or outlet or lateral ditch. The bottom of the field ditch should be on a level with the outlet five to 15 feet back from the junction. Flume channels should be seeded, sodded, or planted. Where grassed flumes will not prevent erosion, a paved flume should be used. The toe of the paved flume should be ten to 15 feet from the bottom of the larger ditch to decrease the danger of silting.

Constructing Ditches. After the ditch system has been planned, stakes are set every 100 feet along straight lines called for by the plan, and 25 to 50 feet along curves. The tops of grade stakes are of a uniform height, usually five feet above the grade of the ditch bottom. These grade stakes are checked by sighting with a level, to be sure their tops line up properly.

As construction progresses, the desired depth of the ditch can be checked by sighting over the top of a gage stick in the bottom of the ditch. In the end, the top of the gage stick should be in

Figure 15-6: Ditch Construction. Note the slope and bottom stakes. What is the purpose of the gage stick?

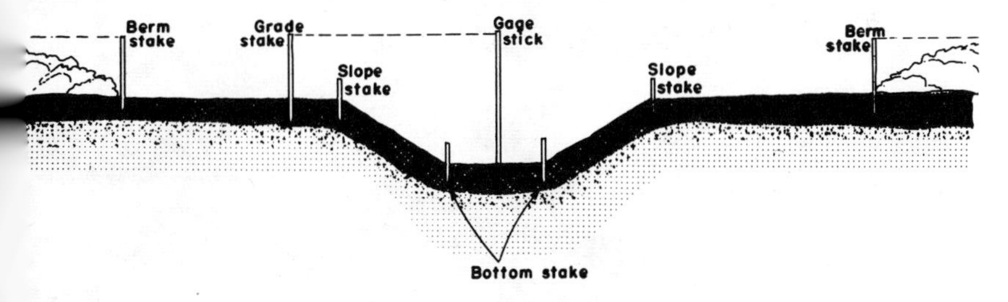

line with the top of the grade stake. Berm stakes show the beginning of the spoil or dump bank.

A variety of machinery may be used for construction of ditches. The dragline is the most versatile piece of large equipment for making all kinds of surface ditches. Pull graders or motor graders are effective dirt movers. An important phase of drainage as well as irrigation is land leveling with graders, scrapers, carry-alls, and bulldozers. Final finishing of land grading is done by the land leveler or planer. Dynamite is used for making ditches in land too soft to support machinery. Special elevator-bucket type equipment is used for making trenches for tile drainage systems, discussed next.

Farming Surface Drainage Systems. Farming operations should be parallel to the field ditches. In plowing, back furrows should be established midway between ditches, and all furrows should be turned toward the middle. This will give the land a considerable crown after a few years and drainage will be improved.

Figure 15-7: Moling. This cross-sectional drawing shows how a panbreaker with a mole-ball attachment forces the soil aside to form a tunnel for field drainage.

Subsurface Drainage

A subsurface or underground drainage system will remove excess soil water. The system is valuable not only for draining flatlands that have high water tables, but for correcting heavy seepage on slopes. When water farther up a slope cannot penetrate the soil and a wet spot develops, it can be drained by placing tile above the wet spot. The tile is buried a few inches in the hard layer of soil along which the water is flowing, and covered with gravel.

Types of Subsurface Systems. Two types of drains are used for subsurface systems, (1) mole drains and (2) tile drains.

Mole Drains. Mole drains are often used in clay, clay loam, and organic soils. They are less successful in peats and mucks.

A moling machine is one that draws a bullet-nosed cylinder, usually six inches in diameter, through the soil. A drainage system four to five inches in diameter is thereby formed. A mole drain should be at least 30 inches below the surface to prevent closing of the holes by compaction from normal farming operations.

Well-constructed mole drains last from five to eight years in soils adapted to their use. When they stop working, new lines are established between the old. How far apart they should be depends on local conditions.

Mole drains are often valuable to supplement open ditches. They help equalize water levels between ditches used for either drainage or subirrigation.

Tile Drains. When properly selected and installed, tile drains give the best permanent type of underground drainage.

Locating Tile Drains. A tile drain system is a series of main drains and laterals. The main drains follow the natural watercourse. The laterals should be straight and run in the general direction of greatest slope. Each main drain should be as long as possible to cut down the number of outlets, which require considerable upkeep.

Depth and Spacing of Tile. The space between tile drains should be narrow enough so that the drainage system will restore an acceptable ground-water level within 24 hours of a rain. Water moves through soils with good tilth quicker than through compact soils. Tiles in sandy soils thus can be placed deeper and farther apart than tiles in clay soils. Tile should always be placed at least 2½ feet deep, to prevent breakage by heavy machinery or by frost.

In clay or clay loam, tile drains are placed 40 to 70 feet apart and 2½ to three feet deep. In silt loam, they are placed 60 to 100 feet apart and from three to four feet deep. In sandy loam, they are placed 100 to 300 feet apart and 3½ to 4½ feet deep. Spacing closer than 50 feet is generally too expensive except for high-value lands.

Size of Tile. Where considerable surface water must be drained by the tile, the drains should be large enough to carry a runoff of one inch in 24 hours. In loam where no surface water will get into the tiles, they should be able to remove three-eighth inch of

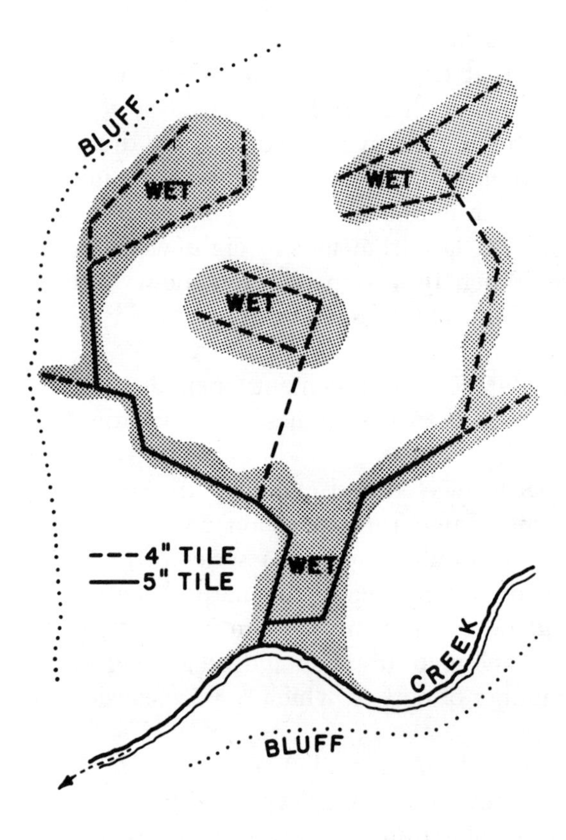

Figure 15-8: Main Drains and Laterals. Before beginning construction of a tile drainage system, good farmers make a map such as this one, which shows a "random" system of tile drainage. Note that main drains follow natural watercourses.

water during the same period. In loose, sandy soil, they should remove one-half inch of water during the period.

A five-inch tile has nearly twice the capacity of a four-inch tile. A four-inch tile is the smallest that should be used in land drainage.

Where surface drainage takes off part of the water, less is dependent upon the size of the tile drains.

Grades for Tile Drains. The minimum grade or slope per 100 feet for four-inch tile is 0.20 foot and the maximum length is 1300 feet. The minimum grade for five-inch tile is 0.10 foot per 100 feet and the maximum length is 2000 feet. The minimum grade for a six-inch tile is 0.07 foot and the maximum length is 3000 feet. Grades or slopes of less than 0.05 foot per 100 feet are not recommended for any size tile. Tile should not have a steeper grade than two feet per 100 feet. Sewer pipe with sealed joints should be used on steeper grades.

Selecting Tile. Water moves by gravity into the joints between tiles, not through the tile walls. Porous tile gives no better drainage than tile that water does not permeate, and porous tile is apt to be easily broken or crushed.

Drain tile should be circular in cross section. The inside should be smooth. The ends should be regular and smooth to permit making close joints of not more than one-eighth inch between tiles. Tile smaller than ten inches in diameter is commonly 12 inches long. Tile larger than ten inches in diameter may be up to 30 inches long.

Good tile is free from visible grain, masses of lime, and minerals that cause slacking or breaking down. A broken surface should show uniform structure throughout.

Common tile is made from brick clay. Concrete tile is difficult to make at home because special equipment and experienced operators are required.

Constructing Tile Drains. Power-operated trenching machines are generally used to dig trenches for tile drains. Machines can do the job more cheaply than can the farmer using hand labor whenever extensive systems need to be installed.

Trench digging should begin at the outlet and proceed upgrade. This permits water to drain from the trench while it is under construction.

Tile should be laid as soon as the trench is graded, to avoid cave-ins and other accidents affecting grade. Special hand tools are used for laying tile. Work should begin at the lower end or outlet. The tiles are turned until they fit well. They should fit tightly at the top if not at the bottom. Cracks or openings of one-fourth inch or more at the top or side should be covered with pieces of broken tile or strips of roofing paper. In some sandy soils, wide cracks should be sealed with mortar.

The tile drain must be laid in straight lines or with smooth curves if it is to remain open. To prevent silting, the flow of water in the tile should be at a uniform rate. Thus changes in grade or abrupt changes in direction must be avoided. There should be no sags in the completed line. Silt will settle in any depression and cause partial or total clogging. The less the fall, the greater the need for accurate work.

Figure 15-9: Trenching. A tile drainage system is being installed with a tile ditching machine owned by the West Virginia State Soil Conservation Committee.

In a wet soft trench, tile may be laid on planks bedded at grade. In fine sand or sandy loam it is desirable to wrap the tile joints with cloth or roofing paper to prevent soil from washing into the tile.

On the day the tile is laid, it should be covered with four to six inches of loose earth to hold it in place. Earth should be packed around the sides. The earth at the joints should not be compacted, however, because compaction might prevent the flow of water into the drainage system. Putting topsoil around the joint may help stop the soil from cementing the tile joints. In tigh soil, gravel should be placed at the joints, and the trench should be filled with gravel to within a foot of the surface.

Protecting the Outlet. Failure to protect the outlet of a tile drainage system commonly causes trouble. Animals may nest or become lodged inside. Washing or eroding of the ditch bank may overexpose the outlet. A reinforced concrete headwall with wing walls and an apron provides good protection. Sewer pipe should be used for the last ten to 15 feet of the drain, or a length of strong metal pipe for the last 15 to 20 feet. Both types should be about two inches larger than the tile discharging through them, so that the pipe can be slipped over the last tile for five or six inches. This joint should be cemented tight to keep out earth. The headwall should project into the ditch or stream, so that the flow from the drain will fall clear of the bank and of ice or debris. If the headwall does not project, the bank should be protected by concrete. If the pipe alone sticks out, it should angle downstream.

A screen or grate should be placed over the outlet to prevent animals from entering.

Silt Wells. In soils where surface water collects in spite of tile drainage, a silt well is often used to carry the surface water into the drainage system. The wells are made of concrete or sewer pipe. They may also serve as junction boxes where two or more tile lines come together.

Where only a little surface water collects, a "blind inlet" may handle it. This is made by filling a small section or trench with stones, broken brush, or gravel. Coarser materials are placed nearest to the tile. In cultivated fields the upper layer should be of porous soil.

Maintenance of Drainage Systems

If ditches are allowed to deteriorate, the cost of repairing them may equal the original cost. A drain can work well only if it is clean. Weeds, briars, willows, silt, trash, and other refuse should be removed. Frequent inspection is a good practice.

A good grass sod on outlet ditches will reduce maintenance costs. Pasturing ditches helps keep down excessive growth. However, livestock should not be allowed on ditches while the banks

Figure 15-10: Grassed Outlet Ditch. Sod like this keeps maintenance costs low.

are soft. Mowing the grass on ditches is another good practice.

A check of how tile drains are working can be made by walking along the tile lines a few days after a heavy rain. Wet spots show that water is not being carried off. A hole or cave-in shows that a tile has been broken or displaced.

Roots of trees growing near tile drains may enter them and obstruct flow. All willows and other water-loving trees growing within 50 feet of tile drains should be destroyed.

Crop rotations and other practices that maintain good soil tilth help keep drains working. Continuous row cropping breaks down tilth to the extent that water may not percolate rapidly enough to the drains.

Summary

Drainage is needed on many level lands to provide plants and soil microorganisms with soil air, to help prevent shallow root systems, to improve soil tilth, to permit early planting and plowing, and to remove excess irrigation water.

There are two kinds of drainage systems, surface and subsurface.

The objective in designing a drainage system is to get the simplest and cheapest series of drains that will allow crops to make satisfactory growth. Plans for small systems may be made by field inspection. For large systems, the entire watershed should be considered, a map should be prepared, and the system should be staked out in the field.

Surface drainage systems remove surface water through open ditches. Field ditches empty into collecting ditches built to follow a natural watercourse. A natural grade or fall is needed, and care must be taken to prevent erosion with drop spillways, flumes, and other devices.

Subsurface or underground drainage systems remove excess soil water from flatlands and seepy hillsides. Mole drains and tile drains are used. Tile drains give the best permanent underground drainage. Tile systems are made up of main drains and laterals. Placements, depths, spacings, sizes, and grades or falls vary. Construction is with power-operated trenching machines. Drains must be laid in straight lines or with smooth curves, even grades, and reasonably snug joints. Outlets must be protected with headwalls and screens or grates.

Ditches must be kept clear if they are to work properly. Tile drains can be checked by walking along the lines after a heavy rain, watching for wet spots or cave-ins.

Study Questions

1. How does drainage help improve soils for both crops and farm operations?
2. What is involved in planning a drainage system?
3. What practices should be followed in constructing surface ditches?
4. How effective are mole drains?

5. How do tile drains provide drainage?
6. What are the functions of main drains and laterals?
7. How are "seepy" hillsides best drained?
8. What practices should be followed in determining depth of tile and spacing of drains?
9. What recommendations are made for grade and length of drains for four-, five-, and six-inch tile?
10. What recommendations are made for selecting tile?
11. What methods are used in constructing tile drains?
12. What recommendations are made for laying tile?
13. What provisions should be made for protecting drainage outlets?
14. What are silt wells and what is their function?
15. List approved practices for maintaining ditch and tile drainage systems.

Class Activities

1. List the drainage problems of home farms and of the community.
2. Arrange field trips to examine surface and underground drainage systems of the community.
3. Examine commercial and government publications on land drainage.
4. Make maps of home farms, showing the natural surface watercourses.
5. Observe the natural watershed areas of the community.

Irrigation Needs and Practices

Anyone who has tried to keep a lawn green through the summer even in the more humid regions of the United States knows that it is often necessary to supplement rainfall and soil water with one form of irrigation, sprinkling. Like lawn grass, all crops need large and well-spaced amounts of water for growth. Like lawn grass, too, most crops do not get enough water at the proper times unless natural supplies are supplemented by irrigation. Water is not only an essential plant nutrient, but a solvent by means of which the nutrients in the soil and in fertilizers are dissolved and made available to plants. Without water, all crop growth stops. Even if water is available, but in amounts less than the maximum a crop can use at any particular time, growth suffers and harvest is reduced.

In dry regions such as the Western states, farmers have long practiced irrigation. With the development of equipment such as aluminum pipe, techniques such as field sprinkling, and professional assistance such as that provided by the Soil Conservation Service, irrigation is becoming more and more common in the East. In this relatively humid region, irrigation is a profitable corrective of poorly spaced rains.

Thus, farmers and ranchers in all parts of the United States are turning to irrigation to increase their crop yields. The increase in yield must pay the cost of additional labor, equipment, ditches, and drains. High-value crops such as sugar beets, potatoes, and truck crops are most likely to do this. In many areas, however, simple irrigation systems can be installed cheaply enough to permit profitable irrigation of all crops.

Types of Irrigation Systems

The three types of irrigation are (1) sprinkler systems, (2) surface systems, and (3) underground systems.

Sprinkler Systems. In sprinkler systems, water is shot into the air with nozzles or perforations in pipes, and falls as an artificial rain.

Surface Systems. Surface irrigation makes use of a series of ditches from which water is spread.

Underground Systems. Underground irrigation is provided by raising or creating a water table so that capillarity will feed roots of crops. This system makes use of deep ditches, often supplemented by mole drains, discussed in Chapter 15.

Combinations. In practice, farmers often use two or more of these systems for irrigation. For instance, ditches may be employed to feed sprinkler systems.

Planning the Irrigation System

Irrigation systems are large investments. Farmers can save themselves money and trouble by taking advantage of professional assistance when making plans for irrigation. Government agencies that will help farmers determine irrigation needs and lay out their systems include the Soil Conservation Service, the Bureau of Reclamation, and Extension and vocational education services. Companies selling irrigation equipment are also generous with advice.

An irrigation system should be designed so that water can be delivered (1) to all fields when needed, and (2) in quantities sufficient to meet moisture demands of crops during peak-use periods. If the water is supplied to different fields on a rotation basis, the delivery system should be large enough to allow the delivery of sufficient water in the time allotted. Irrigation systems must be planned to fit particular soils, slopes, crops, and water sources.

Water Sources. The first requirement for irrigation is a supply of water from streams, lakes, ponds, reservoirs, or wells. If the water source is a stream or a body of still water, such as a lake, that is above the area to be irrigated, gravity flow can be employed to bring the water where it is needed through ditches or pipes. Otherwise, it is necessary to raise the water to field level with pumps.

In some areas of the United States, the only source of irrigation water is the great reservoir of water in the ground. This supply of adequate and suitable ground water is not always available. Alluvial deposits containing thick layers of water-bearing sand and gravel are favorable for storing a good underground water supply. Ground water is often maintained by seepage, in areas crossed by rivers or irrigation canals.

Requirements of Terrain and Crops. The methods used to supply water must fit the land and the crops to be irrigated. A soil survey might be needed to provide necessary information on soil depth, texture, structure, permeability, productivity, and water-holding capacity. The water requirements of different crops also need to be considered. Deep-rooted plants may need less frequent irrigation than shallow-rooted plants.

Figure 16-1: Pumping Ground Water. Well and pumps supply much irrigation water in California.

Figure 16-2: Leveling a Field. Grading is often required for successful surface irrigation.

Land Preparation. The method of water application to b used will determine the degree to which land must be levele and graded. Sprinkler irrigation requires little land preparation For the successful operation of surface and underground sys tems, however, few considerations are more important. Unles low spots are filled, soil will become waterlogged. High spot must be leveled or they will remain dry. Deep surface soils ca be graded more extensively than shallow soils. Leveled spots i shallow soils may not have enough depth for root growth an water retention.

Each of the various surface irrigation systems presents addi tional grading problems, discussed in more detail later in thi chapter.

In states which have extensive irrigation, specific recommenda tions on land preparation are usually issued by agricultural ex periment stations in cooperation with the Soil Conservation Serv ice and other government agencies. These recommendations wi vary for different localities within the state, depending upon th soil and the crops to be grown.

Other Problems. Other problems of irrigation which should be considered in the planning stage are soil erosion, alkali accumulation, and drainage for areas which cannot be filled in. Long-term irrigation may also raise the water table to a dangerous level, and some provision will have to be made to correct or prevent this. Conservation of irrigation water is another factor of importance. Limited or expensive water sources, for instance, may dictate the type of system to be used. Some provision often must be made for the removal of excess water and storm water.

Sprinkler Irrigation

Sprinkler systems provide an excellent control of water. They make it possible to apply water at any rate that can be absorbed by the soil, thus preventing runoff and excessive soil erosion. Sprinklers can be stopped when the soil has absorbed the right amount of water. They can be used on steep slopes where other methods of irrigation are not practical. They are especially useful in establishing pastures on steep slopes. Uniform water application is a further advantage. Sprinkler systems are excellent for soils that absorb water rapidly, because lesser amounts of water can be more uniformly spread than by surface irrigation.

Figure 16-3: Irrigating an Orchard. Sprinklers make it possible to water this established orchard without disturbing roots by installation of ditches or tiles.

They also make it possible to avoid the waste of land which other systems require for ditches.

Sprinkler systems are not suitable for all conditions, however. In hot, windy climates, much water is lost through evaporation. Excessive winds also cause uneven distribution of water.

Sprinkler Equipment. The two types of sprinkler systems are those using sprinkler heads and those using perforated pipe.

Sprinkler Head Systems. Sprinkler head systems are the more widely used. They may be permanent installations with buried main lines and lateral lines, or they may have fixed main lines and portable laterals. The systems also may be entirely portable. Water is discharged into the air through nozzles on the sprinkler heads, which are mounted on riser pipes attached to the laterals.

Figure 16-4: Two-nozzle Revolving Head. Which is the range nozzle? The spreader nozzle?

A variety of sprinkler heads is available. Some are rotating heads and some are stationary. Some have two nozzles and some have one. The most common type is the two-nozzle revolving head. One of these nozzles, designed to throw water farther out is called the *range nozzle;* the other is called the *spreader nozzle*. A jet in the range nozzle causes it to turn at the rate of abou three revolutions per minute. Such sprinklers can apply 0.2 to 1.0 inch of water per hour. The smaller single-nozzle heads are adapted to orchard irrigation but are seldom used for field crops.

The amount of water applied in a circular area is determined by the size of the nozzle and the water pressure. The water is not applied evenly over the entire wetted area, so the system is arranged to overlap. The amount of overlap should be from one fourth to one half of the wetted circle.

Revolving sprinkler heads are manufactured for different amounts of water pressure. Pressure too high or too low for a particular head may cause poor water distribution.

Perforated Pipe Systems. The fixed perforated pipe systems are apt to cause puddling on tight soils that absorb water slowly. Oscillating systems with only one row of holes that apply water more slowly may be safely used on most soils. The width of a strip irrigated by a perforated pipe depends on the size of the holes and the water pressure. The usual systems cover an area 40 feet wide with an operating pressure of 25 pounds to the square inch.

Laterals. Special equipment has been devised for moving portable laterals. Side-roll laterals are mounted on wheels with the pipeline as an axle. Flexible hose is used to connect the main line. Pull-type laterals, mounted on skids or on wheel-carriages, may be used to move the line forward. Crops may be damaged if the lines are pulled across muddy fields.

Whether laterals are portable or fixed, they should run as nearly level as possible to provide uniform water pressure. If this is not possible, they should run down the slope, to offset or partially offset with gravity the pressure loss due to friction.

A source of water near the center of the irrigation area will provide the most economical combination of main and lateral pipe sizes.

Pumps and Power. The equipment of a sprinkler system generally includes pumps to provide needed water pressure. The best pumps are turbine or horizontal centrifugal pumps. The turbine pump is used for relatively deep sources of ground water. Centrifugal pumps are used with surface water or shallow wells, where suction lifts are less than 15 feet. Turbine pumps are generally set in fixed locations.

Figure 16-5: Aluminum Pipe. Different sizes of this handy portable pipe are used with power systems of different capacities.

Electric motors and internal combustion engines are used to drive the pumps. Electric motors are used in fixed locations, while gasoline and diesel engines are adapted for use with portable pumping units. Power units should be equipped with automatic cutoff devices to stop the motor if the main line pressure drops below a fixed safety level.

Booster pumps may be needed to provide adequate water pressure for higher elevations. Take-off valves are generally needed at the laterals.

Pipe sizes should be in balance with the power system for maintenance of water pressure.

Sources and cost of electric power are factors to consider. Single-phase, low-voltage transmission lines will not furnish enough power for large irrigation systems.

Protecting Equipment. Debris screens are needed when surface water is used. Desilting basins may be required to trap sand and suspended silt.

Operating Sprinkler Systems. Water should not be applied faster than the soil will absorb it. Yet it should be applied fast enough to prevent excessive evaporation losses.

Depth of Application. The amount of water applied during an irrigation should not be greater at the point of lightest application than can be held by the soil within the root zone of the crop. Greater amounts of water should be applied only when its purpose is to leach out harmful soil salts.

System Capacity. The equipment should be able to replenish soil moisture in the entire area being irrigated at a rate at least equal to the peak rate of use by the crop.

Uniformity of Application. Water should be applied as uniformly as possible over the field. The point of lightest application should receive at least 80 per cent as much water as the average for the field. Uniformity of application is affected by differences in the discharges of individual sprinklers along one lateral and different laterals. It also is affected by the uniformity of spray distribution.

Crop Protection. Water should be applied in a way that will not physically damage the crop being irrigated.

Surface Irrigation

Types of Surface Irrigation. The most common methods of surface irrigation are (1) furrow irrigation, (2) contour irrigation, (3) broad-furrow irrigation, (4) corrugation irrigation, (5) controlled flooding, (6) border irrigation, (7) contour or bench-border irrigation, and (8) basin irrigation.

Furrow Irrigation. For furrow irrigation, water is run in the furrows between plant rows. Furrow irrigation is the most common method of adding water to row crops. It is suitable for truck crops, orchards, vineyards, and berries. It is well adapted to gentle slopes and to soil with a tendency to bake or crust badly when flood irrigation is used, but not to coarse-textured soil.

Users should be cautioned that deep furrows plus frequent cultivation result in soil and soil nutrient losses through water erosion. While furrows should run directly down the slope, they should be adjusted to reduce flow velocity if erosion occurs.

Figure 16-6: Plowing for Furrow Irrigation. Note the careful grading that has preceded furrowing with this special equipment.

Fields are usually rectangular for this system, and furrows are of approximately equal lengths. A limited water supply, such as that from a well or small pump, can be used most efficiently by the furrow method.

Contour-Furrow Irrigation. Contour-furrow irrigation is a method of applying water in furrows that run across rather than down a slope. These furrows are graded just enough to permit water to flow, but not enough to cause washing of soil. Deep-furrow row crops can be irrigated safely by the contour method on slopes with a grade approaching 8 per cent. The system works best on slopes that are fairly uniform.

Corrugation Irrigation. With the corrugation method, water is applied in small furrows which are close enough together to permit the water to penetrate the area between them. This spreads water uniformly across the surface of the soil. The method is designed especially for heavy soils with a slow water intake, since such soils tend to seal over and bake when flood irrigation is used. It is recommended for close-growing crops such as grain and grasses on steep or rolling land. Extreme care is needed, however, on slopes of more than 2 per cent.

Controlled Flooding. For controlled flooding, water is flooded down a slope between closely spaced field ditches which keep the water from concentrating and causing erosion. Frequent openings in the ditches allow a uniform distribution of water over the field. The technique is suitable for close-growing crops such as small grains, and it is used to establish pastures on rolling land.

Border Irrigation. Border irrigation is another method of controlled flooding. Here, a sheet of water is advanced down a narrow strip between low ridges or border dikes. The method requires that the strip be leveled between the border ridges, and that the grade down the strip be fairly uniform to prevent ponding. The ridges should be low and rounded to permit planting the crop over them. Border irrigation is adapted to hay or grain on uniform slopes of up to 3 per cent, and to established pasture on uniform slopes of up to 6 per cent.

Contour or Bench-Border Irrigation. Contour or bench-border irrigation is similar to border irrigation, but the irrigated strips are laid out across a slope on a controlled grade and the ridges are parallel. The method is best used on fairly uniform, moderate slopes with deep soils that have good water penetration. Large streams can be used safely, and thus less time is required for irrigation. The area between diked borders must be leveled to get uniform water application.

Basin Irrigation. The purpose of basin irrigation is quickly to fill a diked area of level land with water to a desired depth and to allow the water to go into the soil. Basins must be properly graded, and of a size adapted to the soil and the water supply. This method of irrigation is used for close-growing crops. It provides good water control and helps correct alkali conditions.

Design of Surface Irrigation Systems. Water in a surface system is distributed with (1) ditches, (2) headgates, (3) checks, and (4) drops.

Ditches. Ditches, often supplemented with pipelines and flumes, are needed to deliver water to the highest point in each field. The ditches should be laid out to provide the right length of irrigation run for different soils, slopes, and crops. They should be constructed with enough grade for water to flow but not enough to cause soil erosion. On farms where erosion is likely to occur, flumes or chutes, pipelines, and ditch-linings may be needed. These will also help prevent seepage, which may be a serious problem on sandy soils. Pipelines should be designed so that they will not be damaged by high water pressure.

Headgates. Headgates are usually installed where the farm water supply is taken from the main supply ditch. They are constructed of concrete, steel, or wood.

Checks. Checks serve to force all or part of the water in a ditch through turnouts. Checks also are often used to raise or hold a high water level in a ditch, making it possible to operate spiles through the bank or siphon tubes over the bank for row irrigation.

Drops. Drops are used in ditches to slow the velocity of water, protecting the ditch from erosion when it is necessary to run water on a grade steep enough to cause cutting. A pipe or a concrete ditch-lining may serve as a drop. Drops also can be made of wood.

Figure 16-7: Headgate. Note the headwall, sidewalls, concrete apron, and cutoff wall at the outer edge of the apron.

Underground Irrigation

In some areas of the United States, lands are irrigated by water from beneath the ground, in a process called *underground irrigation* or *subirrigation*. For subirrigation to be possible, there must be a barrier in the soil profile to prevent water losses through deep percolation. This barrier may be either a relatively impervious layer in the lower subsoil which permits an artificial water table to be created, or a permanently high natural water table. Above the barrier there must exist a layer of soil permeable enough to permit free and rapid movement of water upward and sideways. An adequate drainage is necessary to maintain a uniform water table, either natural or artificial, under the whole irrigation area.

Water is applied to the soil by underground irrigation systems with deep ditches, which may be supplemented by mole drains. Tile may be used, also. The plants are then fed by

Figure 16-8: Testing Penetration. The soil auger is one of the tools farmers use to determine depth of water penetration.

capillarity. The system has the advantage of cutting surface evaporation and erosion. It generally requires much less wastage of ground than surface irrigation systems.

Operating the Irrigation System

You have already learned some of the factors which must be considered in operating sprinkler and surface irrigation systems. The following topics apply to all systems.

Time to Irrigate. An experienced farmer can tell by examining his soil and crops when he needs to irrigate. Crops show signs of needing water when the leaves turn dark bluish-green. Irrigation farmers usually spade up a six-inch depth of soil to study water needs. If this soil will form a firm ball upon being pressed in the hand, it indicates that enough water is available. Heavy clay soils may need irrigation, however, even though they do make a firm ball.

Soil scientists have developed several instruments for measuring available soil moisture. These have proven helpful in research studies and have possibilities for general farm use.

Amount of Water to Apply. The objective in irrigation is to apply enough water to moisten the root zone without excessive loss from deep percolation (water passing down through the soil to the water table) or evaporation. Water moving into dry soil forms a distinct wetting front. Dry soil is slower to become wet than moist soil. Roots take up water freely from soils moist to their water-holding capacity. Roots grow deeper in soils that are kept moist, but not wet, during the early stages of growth.

Water should be applied for a long enough period to moisten dry soil without appreciable loss from the root zone. In order to bring this about, water must be turned off shortly before the entire root zone becomes wet.

The application of water by sprinkler and surface systems should be slow enough to avoid erosion and water wastage from runoff. Many soils take water at rates between one-fourth inch and one inch per hour. In heavy-textured soils, the water infiltration rate decreases rapidly as the topsoil becomes wet.

Soil and Water Conservation. Improper operation of even well-designed irrigation systems will waste water, damage land,

and reduce production. Excessive use of water may leach needed water-soluble nutrients from the soil. Too-heavy irrigation on higher land may cause waterlogging of rich lower land. Lower ends of furrows may not get the water they need, while excessive water penetration may result at the intake end. Wherever water enters an irrigation system there is danger of erosion.

Surface systems are especially susceptible to erosion. One thing to remember is that increasing the size of a stream in a bare, V-type furrow on steeper grades does not increase the rate at which water enters the soil, but does increase the rate at which water washes soil away. On steeper slopes, the irrigator can usually save both water and soil by using smaller streams. This conservation practice is generally worth the bit of extra time it will require for irrigation. Broad or grass-covered furrows may be used with somewhat larger streams.

Figure 16-9: Lined Ditch. Soil erosion and water seepage may be prevented with concrete ditch linings. Note the turnout in the foreground.

Measurement of Irrigation Water. Farmers need to know both how much water their irrigation systems can provide, and how much their crops require. Farmers participating in an irrigation district and the operators of the district also must make some provision for measurement of irrigation water taken from the main canal. Cooperators in an irrigation district are generally entitled to an agreed amount of water. Even when supplies of water are abundant, farmers need to measure the amount of water they apply to a given field. Over-irrigation is as serious a practice as under-irrigation.

Water Measurement Units. Water is measured in terms of water at rest and water in motion. Water at rest—in reservoirs, tanks, and soil—is measured in specific volume as (1) the gallon, (2) the cubic foot, (3) the acre-foot, and (4) the acre-inch. For water in motion, the rate of flow is expressed in (1) cubic feet per second (cfs), (2) miner's inches, and (3) gallons per minute (gpm).

An *acre-foot* is a volume sufficient to cover an acre one foot deep. It contains 43,560 cubic feet of water.

An *acre-inch* is a volume sufficient to cover an acre one inch deep. It contains one twelfth of an acre-foot, or 3630 cubic feet.

When a volume of water equal to one cubic foot passes a given section of a point of a stream in one second, the flow is

Figure 16-10: Diversion Headgate. Measuring devices are usually installed at the point where the farm system is connected with the main canal.

one cubic foot per second. This is equivalent to a stream one foot wide and one foot deep flowing with a velocity of one foot per second.

The *miner's inch* is the quantity of water that will flow through an opening one inch square in a vertical wall under a pressure head ranging from four to seven inches. The water should have free flow below the orifice or opening, as there must be no back pressure. The miner's inch is defined in terms of cubic feet per second. It has different meanings in different states. One miner's inch is designated as 1/50 of a cubic foot per second in southern California, Idaho, Kansas, New Mexico, North Dakota, South Dakota, Nebraska, and Utah. The miner's inch is equal to 1/40 of a cubic foot per second in northern California, Arizona, Montana, Nevada, and Oregon. In Colorado, 38.4 miner's inches is equal to one cubic foot per second.

When a volume of water equal to one gallon passes a given section of a stream in one minute, the flow is said to be *one gallon per minute.*

The cubic foot per second is recommended as the unit for expressing rate of flow and is generally accepted as the standard. When water is pumped, the gallon per minute is almost universally used.

Simplified Measurement. Among the units for measurement of water there are a good many equivalents that greatly simplify calculations. A few are the following:

1. One cubic foot equals 7.48 (about 7½) gallons (U.S.).
2. One cubic foot per second equals 448.8 (about 450) gallons per minute.
3. One cubic foot per second flowing for 12 hours delivers 0.992 (about one) acre foot.
4. One cubic foot per second equals 40 miner's inches (in some states).
5. In some states one miner's inch equals 11.22 (about 11¼) gallons per minute.

Measuring Devices. The quantity of water that flows through an opening or a channel is equal to the cross-section area of the opening or channel multiplied by the velocity of flow. Translated into a simple formula, this would read $q = av$, or q (quantity of water flow, in cubic feet per second) equals a (area of the cross section of the canal or opening, in square feet) times

v (mean velocity of the water flowing through the opening or the canal, in feet per second).

The simplest device used to measure small supplies of irrigation water is the *weir*. A weir is a rectangular, trapezoidal, or V-notched opening in a ditch. The quantity of water that will flow through the weir in a given time is determined by the height of the water above the bottom of the notch (the "pressure head") and the size or area of the notch.

Figure 16-11: Parshall Flume with Automatic Recorder. Measuring devices help prevent overwatering.

The *orifice*, another measuring device, is a mouthlike opening or vent in a wall built across a stream. The opening is below the water surface. The velocity of water flowing through an opening far below the surface is greater than velocity near the surface.

Thus, the velocity of water escaping from an opening in a wall built across an irrigation ditch can be determined by knowing the height of the water above the opening and the size of the opening.

The *Parshall flume* for measuring water flow was developed at the Colorado Agricultural Experiment Station and named after R. L. Parshall. The smaller flumes are well suited to measuring farm water deliveries. The Parshall flume operates with less loss of head than required for weirs.

The *current meter* is a device used by engineers for measuring flowing water in larger canals and rivers. Calculations are based upon the velocity or rate of flow and the depth and width of the stream.

Figure 16-12: Inspecting Drops and Ditches. Good farmers watch their irrigation systems for signs of soil erosion, excessive water seepage, silting, cave-ins, and undermining of concrete drops such as these.

Maintenance of Irrigation Systems

Once an irrigation system of any kind has been installed, careful inspection and maintenance are needed. Even after the best planning, farmers will find it advisable to make frequent field inspections to determine whether adjustments should be made because of problems that have unexpectedly occurred during the actual operation of the system. Things to watch for are soil erosion, excessive water loss through evaporation or seepage, leaching of water-soluble nutrients, and waterlogging of lower land. Mechanical equipment such as pumps and sprinkler heads must be kept working properly. Checks and intakes and similar devices will need adjustments and repairs.

It should be remembered, too, that conditions will change from year to year and from crop to crop. Appropriate adjustments may be needed in such matters as length of runs and spacing of furrows.

Ditches should be closely watched. Silt must be removed and cave-ins must be repaired. Weeds in ditches may seriously slow the movement of water. Chemical solutions in the stream flow are effective for controlling weeds. Cattails may be controlled by chemical sprays. Chemical weed control is generally cheaper than mechanical control.

Sprinkler equipment should be stored when not in use.

Summary

Most crops will profit from irrigation. Farmers in both West and East use irrigation to increase yields and farm profits. Farmers planning irrigation systems can secure help from a number of government and commercial sources. The first requirement is a source of water. Other considerations are terrain, crops, land preparation, and such problems as erosion and drainage.

For sprinkler irrigation, water is shot into the air with nozzles or perforations in pipes, and falls as an artificial rain. The system provides excellent water and erosion control. Two-nozzle revolving heads are most often used; with these, water is shot into the air through nozzles mounted on riser pipes attached to the laterals. Parts or all of sprinkler systems may be portable.

Surface systems make use of a series of ditches to wet the soil. The most common methods of surface irrigation are furrow irrigation, contour irrigation, broad-furrow irrigation, corrugation irrigation, controlled flooding, border irrigation, contour or bench-border irrigation, and basin irrigation. The surface system is made up of ditches, headgates, checks, and drops.

Underground irrigation or subirrigation is created by raising or creating a water table so that capillarity will feed crop roots. Special soil conditions are required.

Experienced farmers determine when irrigation is necessary by inspecting their crops and soil. Water should be applied so as to wet the root zone only. Improper operation will waste water, damage land, and reduce production. Accurate measurement of irrigation water is essential. Measuring devices include weirs, orifices, Parshall flumes, and current meters.

Careful inspection and maintenance are needed.

Study Questions

1. What are the main methods of irrigation?
2. What government and commercial services are available to aid farmers and ranchers in considering and planning an irrigation system?
3. Why is irrigation becoming popular in relatively humid regions?
4. Describe the sprinkler system of irrigation.
5. Under what conditions are wells and pumps used as a source of irrigation water?
6. What needs to be considered in planning an irrigation system?
7. What needs to be considered in preparing land for irrigation?
8. How can a farmer tell when he needs to irrigate his crops?
9. What considerations determine the amount of water to supply?
10. What are some soil and water conservation practices to consider in surface irrigation?
11. What purposes are served by checks and drops?
12. Describe methods used to measure surface irrigation water.
13. Describe units of measurement of irrigation water.

Class Activities

1. Arrange field trips to study and observe (1) irrigation layouts, (2) equipment, and (3) practices.
2. Study commercial and government publications on irrigation.
3. Arrange for a class or individual student project on measuring flow of water used during irrigation periods to determine efficiency of water use.
4. Observe measure of water flow through existing measuring devices.
5. Have specialists from commercial companies, the Extension Service, and the Soil Conservation Service assist the class study of irrigation practices.
6. Make a survey of the extent of irrigation in the community.

Our National Soil and Water Conservation Problem

17

Soil erosion—the movement of soil particles from one place to another under the influence of water or wind—is a national menace. A vast Federal agency, the Soil Conservation Service, was established in 1935 to check it. Public schools have been urged to provide instruction on the problem, and communities to organize programs for control.

Erosion of the earth's crust has been going on since the first rain fell and the first wind carried dust particles. Erosion has cut out our river valleys and has shaped our hills and mountains. Many of the materials from which soils were made have been worked over several times by processes of erosion. Less desirably, where plant cover of grass and trees was not adequate, badlands and sand dunes were formed. The weird rock shapes of desert areas were carved by wind-borne dust. Coulees (dry gulches) were cut by runoff from the infrequent torrential rains. Identical destructive forces are active today in all areas of the world. Man has aggravated them by destroying or reducing plant cover.

Our natural resources are *soil, water, plants, wildlife,* and *minerals.* These are major sources of national wealth. A conservation program is concerned with all natural resources and their intelligent use. Only minerals are exhaustible; the others can be preserved and improved. For instance, water now wasted can produce electric power, irrigate fields, create industry, sustain cities, foster navigation, and provide recreational facilities.

Forests can be harvested at self-replenishing levels. Wildlife can be encouraged for purposes of insect control and sport. Soils can be made ever more productive.

This and the following chapters deal with conservation of water in addition to that of soil because farmers are well aware that excessive dryness limits crop and livestock production as seriously as does infertility.

Geological Erosion

Geology is the science that deals with the history of the earth. Many geologists now believe that the earth and other planets of our solar system were originally part of the sun, and broke away in the form of clouds of gas when a star passed close by. These clouds gradually condensed and cooled and combined into bare, rocky balls. The moon today may look much as our earth once looked, before it had collected its atmosphere. With the coming of our atmosphere, there came to be water, first in the form of vapor or clouds, then in the form of condensed drops of moisture which fell and collected in low places to form streams, lakes, and seas.

The *topography* or surface appearance of this early earth, including its mountain ranges and valleys and watercourses, has been continuously altered by forces acting deep within the ground. Among these have been volcanoes, earthquakes, slow rising and sinking of vast land masses, and other great movements and pressures. Many areas of our own country were at times covered by vast oceans. Evidences of this are the shells of ocean animals found in soils thousands of miles from the seacoasts.

The earth has also been altered by forces working upon its surface. The effects of these surface forces, by which rock is broken up into soil, are known collectively as *geological erosion.* One of the factors included is the great weathering action of frost, which can split rock into fragments. Gravity pulls rocks from mountainsides, and they shatter far below. Air and water in the ground dissolve minerals. Plants split rock with roots. Wind "sands" rock with tiny dust particles, wearing the softer portions away and leaving level plains far beneath prehistoric

Figure 17-1: The Red River Valley. Homme Dam, in North Dakota, is at the edge of the Red River Valley, a farming region of world-famed richness created by geological erosion when a glacial lake filled with sediment.

plateaus, the original heights of which can be measured by the tops of a few remaining buttes or outcroppings of harder rock. Raindrops falling over thousands of years also wear rock away. Streams carve rock into gullies, canyons, and valleys, and waves pound away coastal cliffs. Glaciers—moving rivers of ice—act as huge bulldozers, leveling mountains, filling valleys, and digging vast lakes, of which our Great Lakes are an example.

When rock fragments created and transported by geological erosion are deposited, new geological features are formed. Wind-transported materials settle into dunes and vast plains. Water-transported materials settle in rich bottom lands and deltas; the Mississippi valley and delta are examples. Much of the soil of the Midwest was deposited by melting glaciers. The rich Red River Valley separating Minnesota and the Dakotas was a glacial lake which filled with sediment.

The United States as the first settlers found it was a product of all these forces. Our major eastern and western mountain systems were created by the forces acting from beneath the ground.

On the surface, geological erosion has been responsible for the coastal flatlands, the great river valleys, the vast glacial prairies and lakes of the Midwest, and the wind-created plains of the West.

Man-Made Erosion

Geological erosion never stops, but it works so slowly in terms of our human time that we speak of land acted upon only by geological erosion as *stabilized.* Before the coming of man, our virgin America was stabilized. The soil was held in place by trees and grass and traversed by clear, seldom-flooding streams, all of which had existed without any apparent change for thousands of years. When man began to cut down the trees, plow up the soil, and graze the plains, the protective cover of the stabilized land was destroyed and the forces of erosion—*man-made erosion*—became uncontrolled. Floods, dust storms, gullies, silted lakes, declining farm productivity, and disappearing wildlife have been the result. Man and only man has created all of the destruction of natural resources that we know as erosion.

A Lesson from History. The Bible refers to lands "flowing with milk and honey" that are now semideserts. Scientists believe that the people living there overcultivated and overgrazed and permitted soil fertility to decline.

Figure 17-2: Irrigation in the West. Irrigation—here being performed by a high-powered nozzle on a machine propelling itself along an irrigation ditch—helps conserve soil in dry regions by maintaining a good ground cover.

An example of a Biblical ghost town is the city of Antioch, founded in 300 B.C. and thought to have had as many as 500,000 inhabitants at its height. It was a major center of Christianity. Some of the Epistles of Paul in the Bible were addressed to the flourishing church there, a few years after the death of Christ. In Antioch has been found a cup thought to have been the one used at the Last Supper. Today, all that remains of this great and famous city is the small Turkish town of Antakya, surrounded by miles of ruins and erosion-destroyed farmlands. What was once a stupendous metropolis has become little more than a village, as a result of misuse of resources.

Our soil scientists believe that what has happened to past civilizations can happen here. We need only to look around our own communities to find evidences of gully erosion and muddy, silt-choked streams and lakes. We as a people need to understand the causes of erosion to appreciate its seriousness.

Causes and Effects of Man-Made Erosion. Man-made erosion has been caused by the following practices:

1. Plowing soil too poor to support cultivated crops.
2. Plowing soil in areas with too little rainfall to support continuous crop production.
3. Breaking up large blocks of land susceptible to erosion.
4. Failure to maintain crop residues on the surface while the soil is not protected by growing crops.
5. Exposing soil on slopes.
6. Removing natural vegetation from forest lands.
7. Reducing and weakening plant growth by overgrazing.

Results of erosion are as follows:

1. Loss of the best part of the soil, the topsoil, with its finer soil particles, better tilth, superior water-retention capacity, more plentiful mineral and organic elements, and helpful bacteria.
2. Reduction of crop yields.
3. Need for greater use of plant and commercial fertilizers.
4. Production of less-nutritious crops.
5. Formation of gullies, by which erosion is speeded and farmland is made impossible to cultivate.
6. Covering of rich bottomlands by soils from poorer highlands.
7. Destruction of roadbanks and removal of bridges.
8. Erosion by stream banks of valuable bottomlands.
9. Silting of ditches, streams, dams, lakes, and reservoirs.

10. Reduction of community income and ability to pay for community services through taxation and business activity.
11. Increase in flood hazards.
12. Waste of water that could be used for farming and other purposes.

Erosion in the United States

The total land area of the United States is almost two billion acres. Over the past 150 years, half of our farm land has been damaged by erosion. The United States Soil Conservation Service estimates that 282 million acres of crop and grazing land have been ruined or heavily damaged. Eroded to some extent are 775 million acres of crop, grazing, forest, and other lands.

Crop lands, which are most susceptible to erosion, are estimated at 652 million acres in the United States. Only 500 million of these are considered suitable for cultivation, and of this

Figure 17-3: America Destroyed. Half of our farmlands have been damaged by erosion; hundreds of millions of acres have been ruined forever.

total about 100 million acres have been ruined or nearly ruined, 100 million more badly damaged, and yet another 100 million left in immediate danger of damage.

The United States Soil Conservation Service estimates soil losses by erosion in our country at five and one-half billion tons annually. Losses from farm lands each year are estimated at three billion tons, or enough to fill a freight train that would girdle the earth 18 times. The equivalent in topsoil of 50 160-acre farms is lost each day. The annual loss in dollars is estimated at nearly four billions. It has been said that if you were to stand on the bank of the Mississippi river near its mouth, you would see the equivalent in topsoil of one 40-acre farm flow past each minute.

The indirect effects of soil erosion cost additional hundreds of millions of dollars. Erosion contributes to floods, damaging highways, railroads, and homes, and causing loss of human lives. Dredging of soil-clogged streams and harbors is a constant expense. Costly dams and levees must be built to hold back excessive water flow. Community sources of water for industrial, personal, and recreational use are often silted and ruined.

Loss of Topsoil. Nature may require 100 to 400 years to create an inch of topsoil, the most productive part of the earth. Topsoil is usually dark in color. This dark color indicates a high organic content, largely the fertile remains of partially decayed plant materials. Topsoil provides the most favorable conditions for plant growth. It includes a high proportion of plant foods and the most available source of plant moisture. It is easily worked. It contains millions of microorganisms, chiefly bacteria, which aid plant growth. Loss of topsoil is the most damaging effect of soil erosion.

Such losses can readily be observed in plowed fields. You can easily see the lighter areas of exposed subsoil, usually on high knobs of ground which are most subject to wind and water erosion. You should also compare the depth of the topsoil in the unplowed areas near fences with the depth of the soil in the cultivated field. You will find the soil of the field to be thinner than that near the fences.

Loss of Soil Fertility. Because of this loss of topsoil, fields are generally less fertile today than they were when first plowed.

Figure 17-4: Waste of Soil, Waste of Water. All of this flood-water and the topsoil it is carrying away could have been held on the land.

Yields have been maintained and increased only by the use of more commercial fertilizers, better crop varieties, and new chemicals controlling weeds, insects, and diseases. Wind and water remove organic matter and the fine clay and silt particles of the soil, all of which are rich in essential plant foods. Left behind are coarse, unproductive sands. Severe erosion removes layers of soil as ruthlessly as might a road grader.

Water runs off this severely eroded soil faster and in greater amounts. It not only carries fine soil particles and dissolved nutrients with it, but itself becomes unavailable for plant use.

In the next chapter you will learn how rapid runoff can be prevented or slowed by terraces, grass waterways, strip cropping, and cover crops. Crop yields are higher on protected soils, and commercial and organic fertilizers are more effective. It should be noted that though crop removal causes a natural loss of soil fertility, loss of plant nutrients from erosion in some areas is estimated to be five times as great.

Waste of Water. Saving our water is of an importance equal to that of saving our soil. American farmers, industrialists, and ordinary citizens are becoming increasingly concerned with dropping water tables, polluted streams, silting reservoirs, and widespread crippling of industry and farming by increasing water demand in combination with decreasing available supplies. It is not only the farmer nowadays who looks at the sky and wishes for rain. It may be the worker and the industrialist from the factory on shortened hours because of limited water supplies, the householder forbidden by law from sprinkling his lawn, the city manager whose wells are becoming polluted by seepage of salt water. Plans for industrial expansion have had to be cancelled because of insufficient water supplies. During the droughts of the 1930's, farmers and ranchers were forced to abandon their lands. Even today, parts of our country suffer from periodic and perhaps needless droughts. Growth of cities in the West is being hampered by water shortages.

Our National Conservation Program

Conservation has always been a concern of many Americans. While it is true that for many years it was possible to move to free, virgin land after a farm had "worn out," few farmers can have lightly abandoned their homes, household goods, and farm

Figure 17-5: The Capitol at Williamsburg, Virginia. Even before America was free, farmers like Washington and Jefferson were experimenting with conservation measures between trips to Williamsburg, where Patrick Henry spoke against the Stamp Act in the Capitol.

buildings, their friends and neighbors, and much of their equipment and stock to start again, with next to nothing, at the backbreaking tasks of cutting trees, pulling stumps, picking rocks, building houses and barns, digging wells, and accumulating anew all the tools of farming and the conveniences of daily life. Both Washington and Jefferson are said to have experimented with conservation measures, and by the early nineteenth century other leaders practiced deep plowing, contour plowing, terracing, and ditching. State governments early introduced laws protecting game and fish. President Grant created the world's first national park, Yellowstone, in 1872. In the same year, a prominent Nebraskan, J. Stirling Morton, introduced Arbor Day. Congress gave the President the right to create National Forests in 1891. The Forest Service became an independent part of the Department of Agriculture in 1905. President Theodore Roosevelt enlarged the National Forests more than all the chief executives before and since, and regulated grazing there for the first time. He called conferences of governors and of official representatives of all North American countries, popularizing conservation throughout the hemisphere.

The National Park Service was established in 1916. The United States and Canada signed the first treaty protecting migratory birds the same year. During the 1930's and 1940's, President Franklin D. Roosevelt introduced the Civilian Conservation Corps, which planted forests, fought insects and plant diseases, and controlled fires. The Tennessee Valley Authority was created in 1933 to preserve and develop a whole region of our country. In 1935 the Soil Conservation Service was created

The Program Today. Today, there is an enormous number o
public and private organizations that promote conservation. O
special interest to vocational agriculture students is the nation
wide system of *soil conservation districts* aided by Federal anc
state governments. Each state has a Soil Conservation Commis
sion or committee which approves the establishment of loca
soil conservation districts governed by local boards of super
visors, directors, or commissioners. Local districts invite the as
sistance of the Soil Conservation Service in making farm con
servation plans. Near each local district is a Soil Conservatio
Service office working under the direction of the state Soil Con
servation Office. The local representative of the SCS is calle

Figure 17-6: Hugh Bennett. Hugh Bennett (1881-1960) was chief of the Soil Conservation Service of the United States Department of Agriculture from 1935 to 1951. A distinguished leader, he received national and international recognition. One of his early booklets, "Soil Erosion—A National Menace," helped establish soil conservation as a national objective.

a *farm planner*. He may be assisted by specialists in soils, engineering, and farm management.

Soil conservation districts are assisted by the state Soil Conservation Service in the loan of land-leveling equipment and similar specialized machinery, and in the procurement of tree seedlings and shrubs for flood-control, windbreaks, tree farming, and wildlife shelter. The Soil Conservation Service assists farmers only through their local soil conservation districts and only when the farmer-member requests such assistance from his district.

In addition to the 3000 soil conservation districts in the United States, other cooperative conservation programs include irrigation districts, water-users' associations, drainage districts, water-conservation districts, and levee districts.

Each of our states regulates conservation districts, discussed above, and protects wildlife. Many pass laws reducing waste of natural resources such as gas. They manage state forests, conserve water, and promote conservation education. The Federal government shares with the states the cost of water conservation, forest fire control, and reforestation. The Federal government assists cities in controlling pollution of streams. Farmers are aided in conservation by trained men and women from the U.S. Extension Service. Agricultural experiment stations are centered at land-grant universities supported by the states.

Figure 17-7: Water at Work. The Grand Coulee Dam is only one of the vast reclamation projects across America. The giant generators of Hoover Dam provide power for prosperity to a huge region. All of the water used in these projects was once wasted.

Our Federal government manages the National Forests through the Forest Service of the Department of Agriculture. States are helped with wildlife conservation by the Fish and Wildlife Service of the Department of the Interior, which also manages refuges and hatcheries. Irrigation and flood control are directed by the Bureau of Reclamation. Flood control along major rivers is the responsibility of the Corps of Engineers of our army. The National Park Service, already mentioned, and the Geological Survey also help preserve or investigate national resources.

Private organizations contribute. Examples are the American Forestry Association, the Izaak Walton League, the Friends of the Land, the National Association of Conservation Districts, and the National Reclamation Association. Among groups of this nature for young people are the 4-H Clubs and the Future Farmers of America.

Another important factor in our national conservation program is the work of business concerns. Privately owned timberlands today are often harvested with selective cutting, by which forests are permitted to replenish themselves. By building dams to produce electric power, private utilities store water and provide recreational facilities. Farm equipment manufacturers have developed many machines needed for conservation practices and have helped publicize them. Some companies are cooperating with the government in developing methods of de-salting sea water. Others are devoting valuable advertising space to public education in preventing forest fires.

Results of Our National Effort. These programs have already had an impact on our national conservation problem. With our great system of dams, we are preventing many floods and impounding water for irrigation, recreation, navigation, and electrical production. Some streams once polluted with topsoil and sewerage are clear again, and teeming with fish. Birds and wild animals once near extinction have been saved. Levees protect many flatlands from floods. New forests, guarded from fire, save soil and rainwater, and provide timber. Whole regions of our country once close to bankruptcy have been restored to prosperity. In the Imperial Valley of California and other regions of the West, water once wasted has made the desert bloom.

Less dramatic but equally important is the aid which this national program has given the American farmer. With Federal, state, and local assistance, he can irrigate and drain, join watershed management districts, build ponds, plant trees, control insects and rodents and plant diseases, and build levees to help prevent floods. Above all, he can learn. The latest information, the latest results of government and private experiments on saving our soil and water for future generations and providing for our own prosperity are available today as never before.

Figure 17-8: "Preserving the Land Is Like Flying Old Glory Full Mast." This prize-winning picture was used by a business concern as the keynote to a national campaign publicizing conservation.

Figure 17-9: America Enriched. Vast areas of our country have been made productive as a result of our national conservation program.

Summary

Soil erosion—the movement of soil particles from one place to another under the influence of water or wind—is a national menace. In addition to soil, our national resources are water, plants, wildlife, and minerals. A conservation program is concerned with all of these. Only minerals can be exhausted. The other national resources can be preserved and improved.

Geological erosion is produced by natural forces working on the surface of the earth. Factors include frost, gravity, soil air and water, plant roots, wind, raindrops, streams, waves, and glaciers. When rock fragments created and transported by these forces are deposited, new geological features such as dunes and plains and bottomlands are created. Geological erosion never stops, but it works so slowly in terms of human time that we speak of land acted upon only by geological erosion as *stabilized*. When man cuts down the trees on stabilized lands, plows up the soil, and grazes the plains, the protective cover is destroyed and *man-made erosion* becomes uncontrolled. Man has created all of the destruction of natural resources that we know as erosion. Many lands once "flowing with milk and honey" have become semideserts as a result of misuse of natural resources. The same thing can happen to America. Over half of our farm land has been damaged by erosion. Of the 652 million acres of crop lands in the United States, 150 million acres are unsuitable to cultivation, 100 million acres have been ruined or nearly ruined, 100 million more badly damaged, and yet another 100 million left in immediate danger. Especially serious are losses of rich topsoil and soil fertility. Our water resources are also menaced, by dropping water tables, polluted streams, silting reservoirs, and increasing water demand in combination with decreasing supplies. Water shortages are already hampering American growth.

Conservation has always been a concern of many Americans. Today, there is an enormous number of public and private organizations that promote conservation. Of special interest are soil conservation districts aided by the states and the Soil Conservation Service. Among private organizations with conservation programs for young people are the 4-H Clubs and the Future Farmers of America. Another important factor in our national conservation program is the work of business concerns. All of these programs have had an impact on our national conservation program. Of great importance is the aid given the American farmer in planning and completing conservation projects. Above all, information on conservation problems and practices is available today as never before.

Study Questions

1. What are our important natural resources?
2. What is meant by *geological erosion?*
3. What is meant by *man-made erosion?*
4. What is the lesson of history concerning soil erosion?
5. How severely have the soils of our country been eroded?
6. How much soil is lost annually in the United States by erosion?
7. What are the kinds of damage from erosion and floods other than soil damage?
8. How important is the loss of topsoil?
9. How important is the loss of soil fertility?

10. How important is water conservation?
11. What are the causes of soil erosion?
12. What are the general kinds of damage caused by soil erosion?
13. What is the work of the Soil Conservation Service?

Class Activities

1. Discuss what has taken place in the community that has caused soil erosion and loss of soil fertility.
2. Discuss what changes have taken place in desirable wild game and fish habitats.
3. Invite the local SCS specialist to tell you about changes which have taken place in the type of farming and in the general productivity of soils and soil tilth of the community.
4. Make a field trip to find places where severe erosion, in such forms as gullies, has happened in relatively recent times.
5. On the same trip, find instances of flood and wind damage.
6. Determine the good as well as the poor practices of farming and ranching in the community which affect soil productivity.
7. Study the effect of soil erosion and depletion on local crop yields.

Soil and Water Conservation and Management on the Farm

18

In addition to participating in Federal, state, and local conservation programs, the individual farmer should study his own farm as a whole to determine which improved soil and water management practices and conservation practices will increase his income and efficiency.

The size of farms needed to make farming profitable has been steadily increasing, and so has the cost of land. Farmers must make use of conservation and soil and water management if they are to receive greater returns from the land they now own. An interesting problem for agriculture students is the selection of methods for doubling farm income on present acreage. This has actually been accomplished by many alert farmers through better management of soil, crops, and livestock. These farmers have found that high yields are economical yields.

Careful *land judging*, discussed in the final chapter of this book, is recommended as an aid in choosing the best methods of soil and water management for a farm as a whole. Land judging, or making a soil survey, involves dividing a farm into areas with similar soil characteristics and crop possibilities, and then choosing the best land treatments and fertility treatments for each. When a farm is divided according to these *land capability classes*, it is easier to decide which farm areas are best suited to crops, to grass, or to trees, where ponds might be profitably constructed, where terraces are needed, which soils might be improved with green manures, and so on.

It has been thought best for study purposes to leave a detailed consideration of land judging till the end of the book, so that you will better understand each aspect of soil science and management before you try to understand application of the subject as a whole to a specific farm or ranch.

The sections in this chapter are organized roughly according to conservation and management methods needed for soils with declining degrees of susceptibility to erosion. Management of the poorest farm areas is treated first, management of the better farm areas is treated last. A discussion of pond-building is added at the end. In the chapter that follows, conservation of soil and water within the areas and during the periods of cultivation is discussed.

Handling Wastelands

Farm wastelands are areas so steep, rocky, infertile, sandy, gullied, or wet that they are not suitable for any farming purpose. The farmer cannot hope to get any income from them. At

Figure 18-1: The Cost of Neglect. Banks of streams that are subject to flooding must be stabilized as carefully as other non-income-producing areas. Willows, underbrush, and rock fill would have prevented this damage. Note the nearly destroyed well at right.

the same time, he cannot afford to neglect them. The most common kind of wasteland, areas so badly eroded that they cannot be used even for tree-farming or pasture, are a constant menace. Gullies can rapidly spread into good fields. Sand and gravel can wash down onto rich bottomlands and ruin them. Runoff may cause serious floods, injuring the farmer's soil, crops, buildings, and access roads. Wind-blown dust may injure crops and machinery.

For these and similar reasons, careful handling of wastelands is needed. Gullies should be checked with dams of concrete, rock, packed earth, or brush. As much as possible, gully bottoms and sides should be seeded with grass or planted with seedlings of trees. Grass and trees should also be established over the whole of the wasteland area to stabilize the soil and lessen runoff. All vegetation must be protected from fire and livestock; some farmers recommend rodent control in addition to fences. Planting should be renewed as needed. Some soil treatment may also be necessary to give grass and seedlings a good start. A soil test will reveal the extent to which liming and fertilization are advisable. Since quantities of these materials are likely to be lost with eroded soil during early stages of wasteland stabilization, heavy applications and repeated applications are often used. Hay and straw mulches are helpful in getting grass and other vegetation started.

Once wastelands have been stabilized, they should be left undisturbed.

Stabilized wastelands often reward the farmer with recreational facilities and provide good homes for beneficial wildlife such as insect-eating birds. Steep slopes and other wastelands well-covered with vegetation help retain and store rainwater and snow. Crops may benefit from this additional source of soil water.

Tree Farming

Timber is a valuable multi-purpose crop. Planted in shelter belts, woodlots, eroded fields, and farm areas unsuitable for cultivation, trees will stabilize soil, hold water, cut wind damage

Figure 18-2: Windbreaks. Note that this careful farmer has chosen tall, narrow trees to provide maximum wind protection for the fields in the background.

to neighboring cultivated areas and crops, and increase the farmer's income through the sale of such products as posts, pulpwood, Christmas trees, and sawlogs. It is important to remember that tree farming is not the same thing as handling wastelands. Trees planted on wastelands should be left undisturbed. Trees that are farmed are intended to be harvested.

Tree farming is becoming more and more popular in many parts of the United States. The first requirement for successful culture is an ample supply of rainfall. Trees are generally selected on the basis of soil texture. Pines are used on sand, hardwoods on loam, and both pines and hardwoods on clay. Various government and private agencies have made studies which will assist the farmer in evaluating sites and choosing trees. It is advisable to prepare the site before planting, with fertilization, liming, plowing, and so forth. Contour planting is recommended. Once trees have become established, thinning is important. Thinning is one aspect of *selective harvest;* as trees are cut at each stage of growth to permit full development of the remaining trees, the cut trees are sold.

If tree plantations are to make satisfactory growth and prevent soil erosion and runoff, they must be protected from overgrazing and fire. It should be noted that the binding effect of roots is not the principal factor in holding soil. Research indicates that the protective influence of underforest litter performs this function. In the unglaciated area of Wisconsin, for example, a four-hour rain of 2.4 inches washed but 17 pounds of soil per acre from an ungrazed woodland, and 745 pounds per acre from a grazed woodlot. A cleared and sodded pasture lost 220 pounds of soil per acre during the same storm. So far as fire is concerned, in southern California burning of forests on steep slopes is followed regularly by flows of mud and debris. One flood of this kind in Los Angeles County carried from 50,000 to 67,000 cubic yards of debris from each square mile of denuded land. The adjacent, unburned land in the same flood-causing storm lost only three cubic yards per square mile.

Other enemies of tree plantations are unwise cutting, insects, plant diseases, sleet, and windstorms. But of all sources of damage, fire and grazing are the most dangerous and most common. Fencing serves to prevent grazing damage. Fire losses may be cut in large plantations with firebreaks—unplanted strips of varying widths. Farmers should be careful of cigarets, matches, and campfires.

In selecting trees for shelter belts, the farmer should keep in mind that low or rounded trees are less satisfactory than tall, pointed trees such as pines. Evergreens are especially useful for trapping snow.

Grassland Farming

The term *grassland farming* sometimes is limited in meaning, so that it refers only to management of permanent pasture. Here, however, it is used in its more general sense. In addition to management of permanent pasture, the definition here includes the use of grasses and similar plants in strips, as permanent soil stabilizers in waterways and headlands, as crops in rotations, as organic fertilizers or green manures, and as cover crops.

Grass is a very effective means of conserving soil and water. It was not realized just how effective such a cover was until the pioneering studies at the University of Missouri revealed the great differences in measured losses of soil and water from land in grass and from land in cultivated crops. These studies showed that only about 11½ per cent of the rain that fell was lost from sod, but that more than 27 per cent was lost from cornland. They also showed that the rate of soil loss was such that it would require only 56 years to erode seven inches of topsoil when the land was in corn continuously, but more than 3500 years with the land in permanent sod.

One of the important ways in which close vegetation protects the ground surface is through the interception of rainfall by leaves. Studies have shown that as much as one-half inch of rain is retained upon the leaves and stems of a good stand of alfalfa. Another important water-conserving effect of close vegetation is that of keeping surface water clear by protecting the soil from the beating action of rain. This is important because it has been shown that a soil absorbs clear water at a much higher rate than muddy water.

Still another effect of grass is the maintenance in the soil of a high content of humus and organic matter. Cultivation is known to destroy organic matter rapidly. Grass cover, whether permanent or in a good crop rotation, tends to maintain a normal or

Figure 18-3: Intercepted Rainfall. This beautiful photo shows how vegetation protects the surface of the soil from beating raindrops and keeps rainwater clear for quick penetration.

Figure 18-4: Renovated Pasture. This limed, fertilized, and reseeded pasture is currently being mowed to provide a thick, close growth of forage. Grazing will not be permitted until sod is well established.

high content of organic matter. A soil high in organic matter may have a rate of water penetration more than double that of other soils. Because of this increased rate of water intake, there is a marked reduction in the amount of runoff.

Management of Grasslands. To obtain the best results from grassland farming, it is desirable (1) to choose grasses and other pasture crops that provide a thorough cover of the ground and a fairly large leaf area to hold water and break the impact of raindrops, (2) to provide conditions favorable to the development of the pasture crop, and (3) to put lands used for a crop rotation into pasture often enough to maintain an adequate supply of organic matter.

Permanent Pasture. In many areas of the United States, permanent pasture is by far the best use for eroded or erosion-susceptible farmlands. Thick, permanent sod stabilizes soil, conserves water, and can produce as many "food units per acre" as feed grains at a much lower cost. To the drier states of the West, sod is of a value for problem areas equal to that of trees in the more humid regions of the East. A fact to remember, however, is that permanent grasslands need as much care as other crop areas. Liming, fertilization, and reseeding are important pasture management practices. Improvement of established grasslands, called *pasture renovation,* pays big dividends. Pasture renovation has increased production of forage four to six times. Ranchers in

Figure 18-5: Grass Waterways. Grass waterways here are used in combination with strip cropping in rotation and on the contour. Waste of soil and waste of water are kept at a minimum.

Texas have reported a 300 per cent increase in beef production from renovated pasture.

Grasses in Strips. Strip cropping is often used on lands with a lower erosion hazard than those requiring woodland or permanent pasture treatment. By this system, strips of sod are alternated with strips of cultivated crops. The topic is discussed in detail later in this chapter.

Grass Waterways. Grass waterways are usually essential to prevent erosion of a strip cropping or terracing system. When a field is laid out for strip cropping, plans should be made for seeding or sodding all large waterways. If the field is in meadow when strip cropping is started, the waterways should be left in sod.

It is usually best to seed or sod the waterways at the same time that meadow seedings are made in the field, but they can be prepared and seeded separately. When the waterway is badly eroded or gullied, it may be necessary to level it with a disk, plow, or grader before preparing the seedbed. A hay or straw mulch will protect the seeding until the grass is established. Later it may be necessary to reseed the spots where the new seeding has been washed out in spite of all precautions.

It is a common mistake to make grassed waterways too narrow. They should be wide enough to mow, and should be clipped at the same time that meadow strips are cut for hay.

Grassed waterways should be protected from all tillage operations. Before crossing a grassed waterway, farmers should raise plows out of the ground, straighten disk harrows, and lift cultivators.

Headlands. Headlands, or turnrows, and field borders should be kept in sod. In many parts of the country, it is customary to cultivate the ends of crop fields and plant four or more rows of intertilled crops or drill small grains at right angles to the regular rows. Since the ends of the fields thus cannot be tilled on the contour, serious erosion may result. Instead of farming headlands, it is advisable to seed them with perennial grasses and legumes and let them remain permanently in sod. They provide a place to turn with equipment and move from one strip to another. They can be mowed with the meadow strips and the grassed waterways.

Grass in Rotations. As suggested above, grass is often included with other crops in a rotation. Grass not only provides a valuable cash or feed crop, but serves to protect soil, save water, and add organic matter to the soil. A number of farmers have experimented successfully with new planting techniques for growing

Figure 18-6: The Value of Grass Rotations and Green Manure. Continuous cropping produced the hard, infertile soil at left. The picture at right shows the same soil restored by 15 years of permanent grasses. Use of grass in rotation and use of green manures will similarly enrich a soil with organic matter.

grass in rotations. Larger yields have been reported from such seeding techniques as planting grass in rows rather than broadcast, in planting grass and clover in separate rows rather than in a mixture, and in planting grass in rows banded with fertilizer.

Green Manure. A green manure crop is any crop plowed back into the soil, whether planted for that purpose or not and whether turned under while still green or after maturity. The function of a green manure crop is to add organic material to the soil. It is similar in effect to the use of grass in rotations, but adds more organic matter because top growth, as well as roots, is added to the soil.

The Value of Green Manure. Green manure crops are used in the expectation that the yields of following crops will be increased. The question might well be asked why the turning under of crops has this effect. The answer is that the addition of decayed plant material increases the content of organic matter in the soil. This results in the physical improvement of the soil and stimulates biological and chemical processes which in the end result in an increase in soil productivity.

Mineral soils well supplied with organic materials are in general the ones on which the largest crops are produced. As soil becomes depleted of organic matter, its ability to produce good crops declines.

Decayed plant material is the source of nearly all of the organic matter in the soil. The roots of plants have undoubtedly been the greatest contributor of organic matter. Other important contributions are made by barnyard manure, green manure, and plant residues such as hay or straw mulches. Not all the plant materials turned under become a part of soil humus, however. Much disappears as carbon dioxide during decay. Thus, no very large addition to soil organic matter can be expected after a single green manure crop is turned under.

Soil in sod is being slowly enriched with organic matter. When sod is first cultivated, the process of decay adds further humus, plus nitrogen and minerals. The longer cultivation is continued, however, the more these beneficial elements are depleted.

This loss is not to be altogether deplored, being in large part merely the necessary accompaniment of making organic matter available to crops. The fact must be recognized, however, that if it is allowed to go too far, declining crop yields will result. The use of farm land should be so planned as to maintain the organic matter with grass rotations and green manures, as far as is consistent with a profitable use of the soil. Naturally, if soil organic content can be maintained with grass in crop rotations alone, these are much to be preferred to the use of green manures. Grassland or pasture farming gives the farmer a cash income,

Figure 18-7: Sod-Seeding. The sod-seeder is used to plant and fertilize a forage crop in bands on the contour with little disturbance of existing sod or of soil-conserving standing stubble. The remarkable results that have been achieved by this new technique are suggested in the other picture. Farmers report almost total prevention of soil and water losses, and pasture ready for grazing months earlier. Will this be the farming of the future?

while green manuring ties up a field for a year during which no income can be expected. Pasture renovation, described above, may be much more profitable than the use of green manure to improve the yield of feed grains.

Cover Crops. A cover crop is any crop planted primarily to protect the surface of the soil. A cover crop may or may not later be plowed under. Cover crops are used to prevent erosion and leaching, to shade the ground, and to protect it from excessive freezing and thawing. This last is often of special importance in orchards. Because they tend to hold soil and water and to add organic matter, cover crops have beneficial effects like those of the uses of grasses already discussed. The difference is that cover crops are not "crops," in the sense of something raised for harvest.

In addition to protecting the soil of orchards, cover crops are often planted on cultivated fields in the late summer or early autumn. They then make enough growth to hold soil and snow over the winter. In the spring they are plowed under in preparation for regular planting operations.

Sod crops can occasionally be interseeded with other crops. When a sod crop is suffering from nitrogen deficiency, for instance, a legume crop can frequently be seeded into the sod with moderate success. It must be kept in mind that the seedlings of the new crop have to meet the competition of established plants and consequently must have some advantage over the competing crop in order to succeed. It may be possible to establish a temporary crop during a season when a sod crop is dormant.

Terraces

Terraces are conservation and management devices used on lands too steep to permit strip cropping, but not so steep that permanent pasture or woodland must be used. Another kind of terrace is used to hold water on dry flatlands.

Terraces are earth ridges, usually 18 inches in height. Their basic function is to intercept water, which is either absorbed or conducted slowly from a field. There are two kinds, the level terrace and the drainage or grade terrace.

Figure 18-8: Another Kind of Level Terrace. This Western pasture has been ridged on the contour to hold limited rainfall and snow on the land.

The *level terrace* is a ridge built on sandy soils with little or no grade. It is designed to hold water in the field until absorbed. These terraces are adapted to any area where the rainfall and soil characteristics are such that there is only slight danger of water accumulating and breaking them. They are constructed by moving earth from both sides to form the ridge wall.

The *drainage terrace* consists mainly of a channel across a slope. It is constructed by moving earth from the slope downhill to form a low, flat ridge. The grade of the channel is variable, being level to nearly level at the upper end and increasing little by little along the length of the terrace to afford increased water capacity without increased terrace width. The grade and shape of the channel are so formed as not to produce too fast a runoff of water. In fact, the velocity should be low enough to allow deposition or settling of soil washed from above.

Terrace Design. The cross sections of both level and drainage terraces must be such that available tillage equipment is readily adaptable to working on the side slopes of both the ridges and the channels. The capacity of a drainage terrace should be large enough to conduct safely from the field the runoff from a rainfall of the maximum intensity to be expected during a ten-year period. The water-damming capacity of the level terrace must also be great enough to prevent breaking of the terrace ridge by the water accumulated on the field during a rainfall of the same

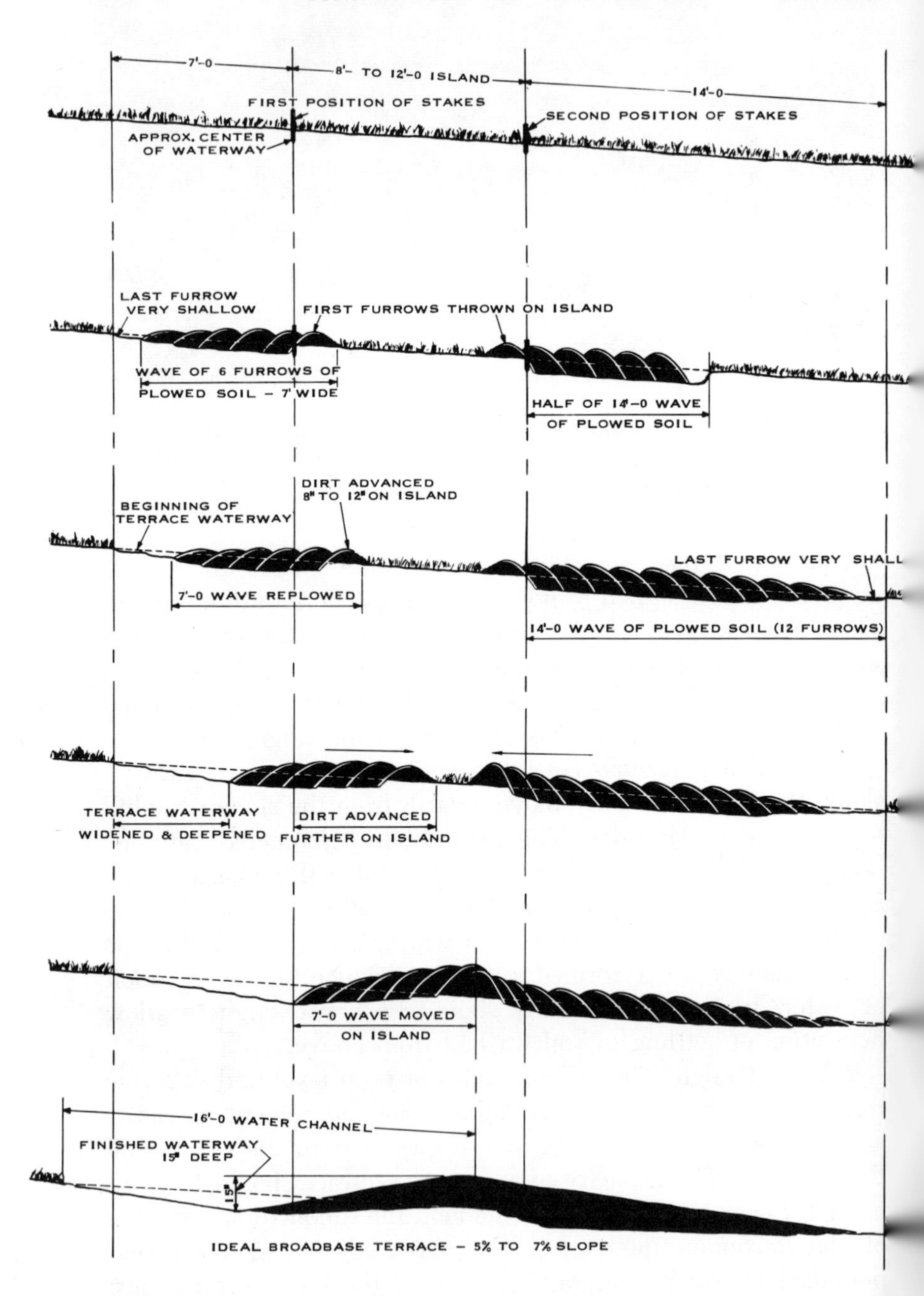

Figure 18-9: Building Terraces with the Plow. The diagram shows the steps of terrace construction with the plow. Note the width of the finished channel and the gradual slopes of the terrace banks.

maximum intensity. The cost of construction, however, often prevents the building of all terraces in dimensions large enough to do this.

Terracing Machinery. A wide variety of machinery is used for terrace construction. Included are the moldboard plow, the disk plow, rotary tiller, elevating grader, and blade grader. The blade grader has been used after adaptation to heavy tractors and even to road patrols.

Under some conditions, farmers can use the plow, disk, and small blade grader to build satisfactory terraces with farm power. The process is difficult, and in heavy soils few good terraces have resulted from the use of this equipment. The larger, power-operated, reversible-blade grader is generally more economical for constructing drainage terraces.

Terrace Location. Spacing or location is one of the most important considerations of terracing. The spacings most often recommended at present are largely the result of experience. Many terrace systems have been studied by engineers, who have used the average spacings of those giving the best results as a basis for spacing tables.

The spacing of level terraces is usually determined by the damming capacity needed to hold maximum rainfalls of an area.

The spacing unit most often used for the drainage terrace system is the *vertical interval,* that is, the difference in elevation of a point on one terrace to a corresponding point on the next. This interval varies with different slopes, being greater on steeper slopes. The increase is not as great as the increase in slope, however, and on flatter slopes the horizontal distance between terraces will therefore be greater. The determining factors are channel capacity, erosion between terraces, soil characteristics, and interference with tillage operations.

Terraces may be spaced close together on steep slopes to prevent concentration of water-washed rills between them. Interference with farming operations, however, encourages wider spacing. Usually spacing and dimensions are used that will give reasonable security from erosion.

Soil characteristics influence the extent of erosion between terraces. The absorption power of a soil will affect the total amount of runoff and erosion. Terraces should have wider spacing on the more absorptive soils.

A final factor to consider in location of terraces is the future use of the land. Land to be put into permanent pasture or other close-growing crops will seldom require the additional protection of mechanical control with terracing. The exceptions to this rule are lands so damaged by erosion that mechanical protection such as that supplied by terraces is necessary before a protective vegetative cover can be established.

Maintaining Terraces. Maintenance is necessary if the terrace system is to continue working properly. Deposits of eroded material in the channel of a drainage terrace will raise its water-flow line and decrease its water-carrying capacity. Tillage will wear down the ridge to some extent, in the cases of both level and drainage terraces. Settlement of the terrace after construction also reduces water-carrying capacity of drainage terraces and water-holding capacity of level terraces.

Once low places and breaks in terraces have been mended, terraces can be successfully maintained by regular plowing operations. To maintain the channel of the drainage terrace, the first plow furrow is started at the outside and turned uphill, leaving the water furrow, or joining cuts of the plow, at the channel flow line. One or two plowings in this manner will keep the channel clear and, if necessary, increase its size by widening and deepening. To maintain ridges of both drainage and level terraces, a back furrow is made at the ridge top and successive furrows are thrown toward it until the entire width of the ridge has been covered. Thus, the height is increased and, if desired, the width also.

Waterways and Headlands. Grassed waterways and headlands, treated above, are usually essential with drainage terracing systems.

Strip Cropping

Strip cropping, you remember, is a variation of grassland farming, by which strips of sod or other close-growing crops are alternated with strips of cultivated crops. Strip cropping is often supplemented as an erosion-control and management device with crop rotation and contour planting. The success of the strip cropping system depends on the ability of the grass or hay strips to reduce the velocity or speed of wind and of water runoff, and to filter out the soil eroded from the corn and grain strips. A 50-foot strip of good alfalfa will filter out 80 per cent of the soil washed from corn or grain strips.

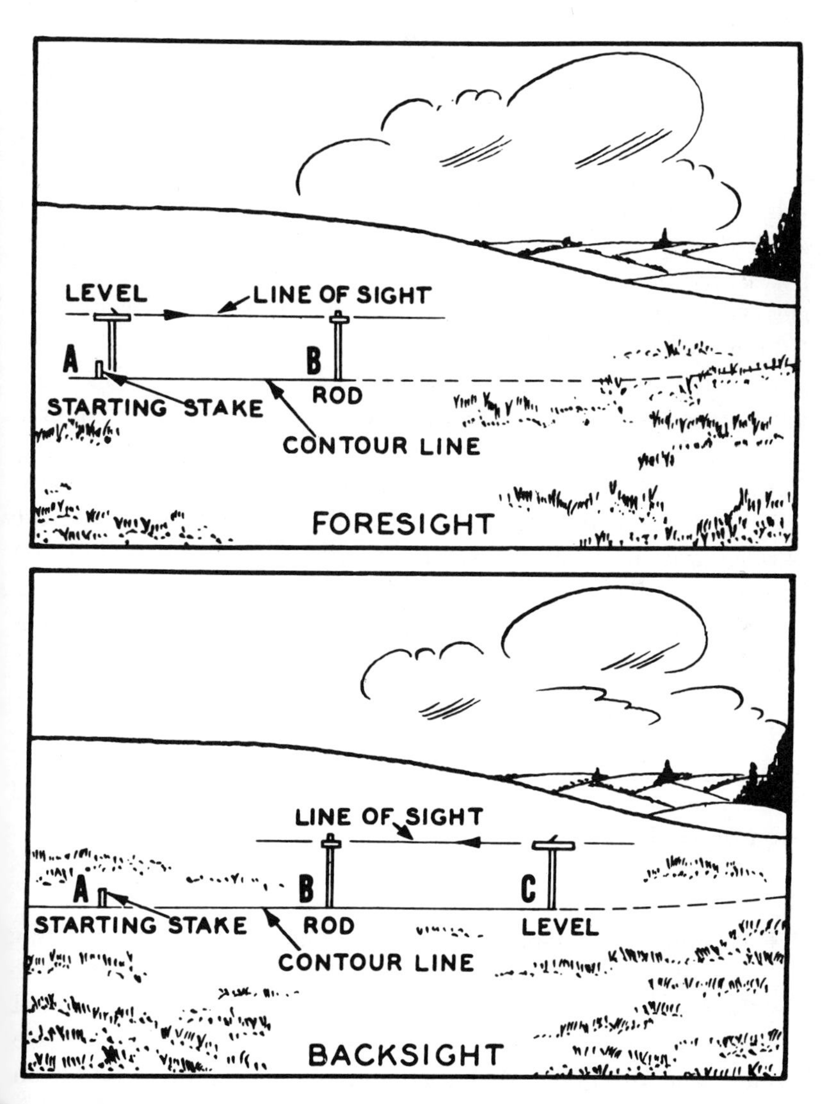

Figure 18-10: Laying Out Contour Lines. Note the simple equipment used for this important farming task.

Types of Strip Cropping. Four types of strip cropping are (1) contour strip cropping, (2) field strip cropping, (3) wind strip cropping, and (4) buffer strip cropping.

Contour Strip Cropping. A *contour line* is a level line laid out across a slope. In contour strip cropping, crops are arranged in strips of identical or variable widths on the contour at right angles to the natural slope of the land. Usually, all the strips are cropped in a rotation, although all of the crops in the rotation need not be in the same field or field unit in the same year. This type of strip cropping is used generally for the control of water erosion. However, it is also effective on sloping lands in areas subject to wind erosion.

Field Strip Cropping. In field strip cropping, or *block farming,* the strips are parallel and of uniform width. They are placed across the general slope, but they do not curve to conform to any contour. The system is recommended only in areas where the land is too irregular or undulating to make contour strip cropping practical. It is a poor substitute for contour strip cropping, but is far better than exposing to erosion entire fields with long slopes.

Wind Strip Cropping. In wind strip cropping, the strips are parallel and uniform in width, usually straight, and laid out across the direction of the prevailing winds. Width is adjusted to assure efficient use of farm machinery. This system is recommended for level or nearly level land where erosion by water is unimportant. It has very little value in conserving moisture.

Buffer Strip Cropping. In buffer strip cropping, permanent strips of grass or legume crops are laid out between strips of crops grown in a rotation. The strips may be wide or narrow, and even or variable in width. They may be placed on steep, badly eroded areas of a slope, or they may be used at more-or-less regular intervals on the slope.

Choosing a Strip Cropping System. The kind of strip cropping to be used depends on a number of local factors. Among these are the kinds of crops that can be grown, the type of rotation to be followed, the kind of erosion (wind or water) to be prevented, and the physical characteristics of the soil.

Laying Out Contour Strips. There are many ways of laying out contour strips. Three general methods will be discussed (1) laying out both edges of the strips on the contour lines

(2) laying out sets of even-width strips measured from a "key," or "base," contour line, and (3) laying out alternate even- and irregular-width strips. Regardless of the method used, one or more contour lines must be laid out on each field to be strip cropped.

A contour line, again, is simply a level line across a slope. With a little experience in the use of a surveyor's level, a farm level, an inexpensive hand level, or even a carpenter's level, it is easy to stake one out. Different operators have different approaches. Some start in the middle of the slope and lay out the longest strips, leaving until last the odds and ends at the top and bottom. Others start at the top, still others at the bottom.

Because of the variation in methods due to local conditions, it is always advisable to obtain the assistance of the SCS farm planner or someone who has had experience in laying out contour strip cropping systems. The effectiveness of strip cropping depends on how closely the strips follow the contour, but the convenience of farming depends on the length and shape of the strips. The more uniform the strips and the smaller the correction areas (areas that depart from the even widths needed for the convenient operation of most farm machinery), the more likely it is that the farmer will follow them.

Figure 18-11: Location of Point Rows. When both sides of the strip follow the contour, most farmers plant from each side toward the center. What are the advantages of this practice?

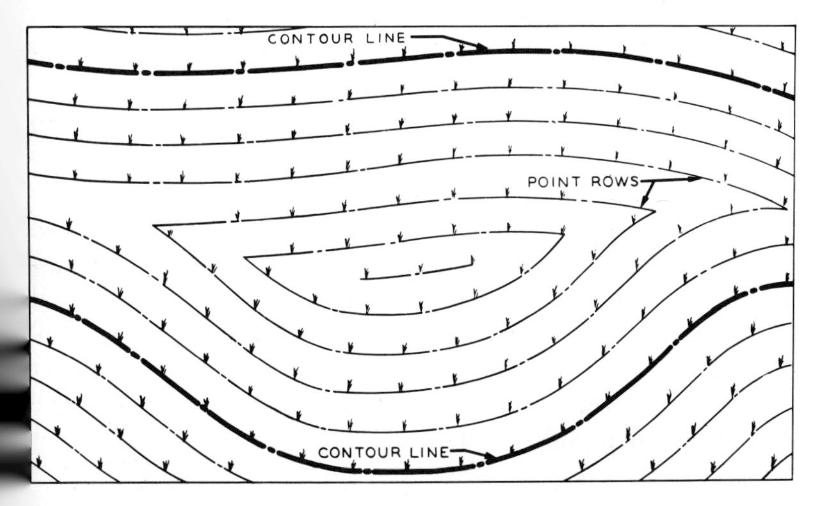

Both Edges on the Contour. When the strip cropping system is laid out so that both the upper and lower edges of the strips are on the contour, all strips are irregular in width. On fairly uniform slopes, the width of the strips thus will be fairly uniform; where the slope varies widely, a great deal of irregularity will result. Uneven-width strips necessitate the use of short and point rows, but all farming operations are on the contour, which is most effective in preventing runoff and soil losses. Some farmers decrease the disadvantage of irregular strips by planting the odd areas with perennial hay crops.

Other farmers make minor adjustments in the locations of the irregular strips in relation to the contour lines. With this system, a minimum of land is devoted to correction areas, and correction areas may be located on the poorer land.

The year the strips are planted to cultivated crops, the number of point rows can be cut by planting them with a close-growing annual crop.

Sets of Even-Width Strips. When the difference in the width of the strips varies too much, operators may prefer to lay out two or more even-width strips from one "base" contour, and use an irregular-width strip between each set of the even-width strips. The irregular-width strips are called *correction strips.* They may be used as buffer strips, or in the regular rotation. This method provides many even-width strips for farming without point rows, but all farming is not on a true contour.

Alternate Even and Irregular Strips. The third method provides for alternate strips of even and irregular widths; it is distinct from the second method because in the second method the even strips are multiple rather than single.

To lay out alternate even and irregular strips, an even-width strip is measured above the first contour line, which may be staked from any point in the field. The second contour line is staked at a distance equal to the width of two strips either above or below the first contour line. An even-width strip is then measured out above this second contour line. This process is repeated until the entire field has been laid out in a strip cropping system. When this method is followed, each contour line forms the lower boundary of an even-width strip and the upper boundary of the adjoining irregular-width strip.

Width of Strips. There is no rule for determining the width of strips that will apply under all conditions. Factors that determine width of strips include (1) degree and length of slope, (2) permeability of soil, (3) the soil's susceptibility to erosion, (4) the amount and intensity of wind and rainfall, (5) kinds and arrangement of crops in the rotation, and (6) the size of farm equipment. In humid areas, strips from 50 to 150 feet in width are in use. In dry areas, strips are frequently eight to 12 rods wide, though others may be as narrow as 12 feet. In general, strips should be of a width convenient to farm, yet not so wide as to permit excessive soil and water losses.

It has been found that more soil and water losses occur when the strips are too far off the contour than when the strips of tilled crops are too wide. On uneven slopes where strips do not follow the contour closely, they should be narrower than on uniform slopes with only slight variations from the contour. Strip cropping does not reduce the length of slope; if the watershed is too long, soil losses will still be high.

Tillage Practices in Strip Cropping. When a strip cropping system has been carefully planned and laid out, it can be maintained permanently by proper tillage practices. In general, operations are no different for farming strips than for whole fields, but some extra precautions are needed.

The crop rotation should be maintained and followed. Meadows should be plowed and planted to the proper crops according to plan. If meadow seedings fail on one series of strips or fields, they should be reseeded so as not to interfere with the order of crops in the rotation or in the strip cropping system.

The plowing of strips should be varied so as not to build up high ridges on the edges and deep dead furrows in the centers. The equipment used and the shape and the location of the strips may determine how the strips should be plowed. If a two-way turning plow is used, all land can be turned up the slope so that water can seep under the furrow slice. After a little experience, strips can be plowed quite easily.

When planting contour strips of uniform width, start at either the top or bottom and continue across the entire length of the strip. Where the strips are irregular in width and both sides are on the contour, most farmers prefer to plant from both sides toward the center. This practice places the greatest number of rows on the contour with the point rows all in the center of the strip, and is the most effective row arrangement to conserve both soil and moisture. Many farmers feel that it also saves time by eliminating long deadhead trips in cultivating and harvesting long and short rows together. Thus, corn can be cut for silage or husked, by hand or with machinery, by starting on the long outside rows, leaving the short point rows to be harvested last and eliminating the need of turning on the crops of adjoining strips.

Figure 18-12: Rotation in Contour Strips. Farmers find that all crops make better growth under careful management such as this.

Crop Rotation

Crop rotation is the growing of a selected number of different kinds of crops in regular order on any particular field. The principal objectives of a good rotation are to secure more economical and more certain production of crops over a period of years, and to control soil erosion.

The Value of Crop Rotation. An historic piece of ground to all interested in agriculture is that of the Morrow Demonstration Plots, located in the center of the University of Illinois campus at Champaign and Urbana. These plots were established in 1876 to demonstrate the long-term effects of different cropping systems with and without soil treatment. The Morrow Plots all are planted to corn every six years, permitting comparison of yields. Yields of the plots planted to continuous corn decreased to 23 bushels per acre by 1920. The plots with a rotation plus treatment with manure, lime, and phosphate were yielding 100 bushels per acre.

A study made at the Missouri Agricultural Experiment Station gave the following results, showing how a good crop rotation can reduce soil losses.

Results of the Missouri Experiment

Cropping System and Cultural Treatment	*Average Annual Soil Loss Per Acre*	*Percent of Rainfall Runoff*
Bare, cultivated, no crop	41.0 tons	30%
Continuous corn	19.7 tons	29%
Continuous wheat	10.1 tons	23%
Rotation: corn, wheat, clover	2.7 tons	14%
Continuous bluegrass	0.3 tons	12%

The prevention of loss of water by runoff, as shown in this experiment, is of special significance for areas of limited rainfall or those subject to occasional drought. Where the crop rotation was practiced, the loss of water was less than half the amount lost from bare land and from land in continuous cultivation, and only one sixth more than from bluegrass sod. A well-chosen rotation keeps the soil in good physical condition, allowing rapid entry of water from rain. Soils that take water readily have less water accumulating on their surfaces during rainstorms. Thus runoff and soil erosion are both decreased.

Values of crop rotation include (1) keeping the soil in suitable physical condition, (2) helping to maintain the supply of organic matter and nitrogen in the soil, (3) providing a practical means of utilizing farm manure and fertilizer, (4) keeping the soil occupied with crops, (5) changing the location of the feeding range of crops, (6) counteracting the possible development of toxic substances in the soil, and (7) improving crop quality.

Keeping the Soil in Suitable Physical Condition. The rapid loss of organic matter resulting from the continuous growing of one crop has a distinctly bad effect on tilth. Loss of organic matter from sandy soils has a tendency to make them looser and drier and subject to blowing and leaching. Other soils are inclined to puddle as organic matter is removed. Rotations that include such crops as grass and deep-rooted legumes will help prevent these conditions.

Maintaining Organic Matter. In addition to improving the physical condition of soil, organic matter is an important source of plant nutrients. Nitrogen is generally present in soils in direct proportion to the amount of organic matter they contain. Organic matter is also needed as food for soil microorganisms, which convert many plant nutrients into forms usable by plants.

Utilizing Manure and Fertilizer. Crop rotation makes it possible to apply manure and fertilizer (and lime) to the most responsive crops, or to those crops that have a long growing season or a high money value.

Keeping the Soil Occupied. Crop rotation makes it possible to have the land occupied with profit-producing crops most of the year. In addition, the loss of plant food by leaching is minimized and losses from erosion are greatly reduced. A crop that leave the soil resistant to erosion may precede a crop that provide little soil protection.

Changing the Feeding Range. When shallow-rooted crop (grasses) and deep-rooted crops (legumes) are alternated, different soil areas serve as feeding ranges, thus doubling the amour of available plant food. An additional benefit is that deep-roote plants in alternation with shallow-rooted crops improve th

Figure 18-13: Exposing the Soil. A rotation suitable to a rich, level field may be dangerous on a steeper field. Which crops are most likely to be unsuitable to cropping on slopes?

physical condition of topsoil and subsoil. Deep-root penetration provides better drainage, because channels are left after the roots decay. Deep roots also draw food elements from lower depths and leave root residues deep below to add to the organic matter and plant food content of the subsoil.

Counteracting Toxic Substances. One crop grown year after year may lead to the development of certain poisons injurious to plants, because of improper decomposition of organic matter. Crop rotations aid the normal decomposition process and thereby prevent the formation and accumulation of toxic substances in the soil.

Improving Crop Quality. Crop rotation not only favors higher crop production, but as a rule insures better crop quality. For instance, quality factors of grain—including test weight, plumpness of kernel, and protein content—are generally improved as the result of crop rotation.

Factors Determining the Rotation. The type of crop rotation to be used on a farm involves consideration of many factors. Among these are the fertility of the soil, and its topography, tilth, drainage, and slope; annual rainfall; prevalence of various weeds, diseases, and insects; and the length of the growing season. Rotations suited to level land, for example, may be entirely unsuited to steep slopes on the same farm.

Erosion and Crop Selection. Erosion hazards are an important factor in determining the kind and sequence of crops to be grown in rotation on a particular piece of land. The amount of erosion is influenced by the texture, structure, and organic content of the soil. On lands subject to excessive erosion as a result of either the slope or the character of the soil, cultivated crops such as corn or cotton are hazardous even when grown in rotation. Deep and fertile lands are adapted to such crops only where the slope is relatively slight. As the slope increases, cultivated crops should give way more and more to the small grains. Rotations will not provide erosion protection on steep slopes. Such slopes need to be protected by permanent pasture or woodlands.

Farm Ponds and Dams

Artificial ponds are becoming ever more popular on farms and ranches. Hundreds of thousands of additional ponds are needed for soil and water conservation and better farm management. Ponds are valuable for livestock watering, fish-cropping, irrigation, fire protection, and recreation. They protect lands below the dams from floods and washing. They allow silt in suspension to settle, clearing water downstream and helping to protect irrigation equipment. They hold or *impound* water which otherwise would be lost as runoff; this water may benefit the farmer by entering the soil or raising the water table. Silted ponds may provide the farmer with rich, deep fields.

In many instances, small ponds can be built by farmers themselves, with scoops or other dirt-moving equipment, during light-work periods. Large ponds generally will require hired construction crews. In either case, it is advisable first to secure the aid of men having a knowledge of surveying and pond construction. Men having such training are members of the Soil Conservation Service, the state conservation department, or the state Extension engineers, county agents, and county engineers.

Figure 18-14: Wasteland at Work. This farmer has dammed a gully and stocked the resulting half-acre pond with bass and bluegills. Note that livestock has been fenced out.

Pond Location. The terrain largely determines the location of the pond, but the use for which it is intended must also be taken into consideration. Different sites will be desirable for livestock watering, fish-cropping, irrigation, fire protection, or recreation.

Once the general area has been determined, the agent or engineer will try to find a small, gradually sloping draw or valley having relatively steep sides and an adequate supply of water. In some cases, a small gully may offer a good location for a pond. The most common sources of water are surface runoff from a well-vegetated drainage area, and springs. The size of the drainage area is of utmost importance. Excess water from too large an area shortens the life of a pond and has a bad effect upon fish management. This is due to silting and to overtaxing the capacity of the spillway.

Pond Size. A pond should be large enough to meet the need for which it is constructed. Naturally, the size will be limited by depth, height of the spillway, location, capacity, source of water, and amount of seepage and evaporation. The most common use of ponds is for livestock water. For this purpose the amount of water impounded should be about twice the quantity actually

Figure 18-15: Dam Construction. Bulldozers are often used to move dirt during construction of a pond.

needed. To maintain this supply, the following depths are suggested in relation to the annual precipitation:

Annual Rainfall	*Required Depth*
40 inches or more	8 feet
20 to 40 inches	8 to 10 feet
20 inches or less	10 to 16 feet

The Spillway. Determining the size of the spillway must not be left to guesswork. Exact dimensions should be figured out. Many pond failures have resulted from spillways of insufficient size and improper construction. If the pond is to be stocked with fish, it is important to have a spillway of sufficient width to cause a shallow overflow.

For small ponds located below watersheds of from ten to 15 acres or less, natural or constructed sod waterways are desirable. Ponds below larger watersheds often require spillway structures. In designing and building these, most farmers will need technical assistance.

After the surveyor has thus completed his work, construction stakes showing the extent of earth fill, the waterline, and the spillway can be established.

Construction. The first requirement of actual construction is the removal of all brush, trees, stumps, stones, and other debris. The dam area can be plowed or scarified to insure a good bond between the original earth and the fill for the dam. For stock-watering purposes, it is important to install a pipe, below the proposed fill, leading to a tank. Normally, not less than 1¼-inch pipe is recommended for this purpose. The stock water-supply pipe and tank should have suitable valves to control the water from the pond. If, to harvest fish or for some other reason, it is necessary to drain the pond at intervals, the larger the pipe the more quickly the pond can be emptied. Where expense is not a factor, a large drain pipe fitted with the proper valves will be found most satisfactory.

In order to prevent loss of water through seepage, a trench should be dug across the center of the dam site and packed with clay or some other heavy earth to form an impervious core. This core should not stop at the original ground level but should be built to a height a little above the proposed water level.

Dirt for the fill is usually scooped from the pond basin and the spillway area. During the digging, the pond can be shaped. Particular attention should be given to the edges and to points of water entrance. For weed control and fish management, the edges of the pond are generally dug down to provide at least a two-foot water depth. It is better to have fewer water entrances, as this allows the water to collect before flowing into the pond. When these entrances are seeded to a dense water-tolerant vegetation, they will strain out sediment coming down from the watershed and reduce the amount of silt carried into the pond.

A good earth fill dam can be made with a three-to-one slope on the water side and two-to-one slope on the downstream side, with the fill dirt placed on a level in six- to eight-inch layers. Packing by earth-moving equipment is generally sufficient.

The top of the dam should be two to three feet higher than the bottom of the sodded spillway. The width of the top is determined, to a large extent, by the type of equipment used in construction, but usually at least an eight-foot top is desirable. When the top of the dam is to be used as a road, a 12-foot top will be found more suitable. Building small dams one foot above the planned height is usually sufficient to allow for settling.

To protect earth dams against erosion, they should be sodded or seeded with soil-holding and binding plants adapted to the area. Generally, erosion of the fill by wave action is not a factor with small ponds. If this condition does occur, it can usually be controlled by planting water-tolerant grasses, reeds, and other plants along the waterline. Under some conditions it may be desirable to riprap this area with stone or a similar material.

Keeping livestock away from the pond itself by fencing generally is advisable. This reduces maintenance, provides clean water for stock, and permits fish production.

Summary

With the aid of careful land judging or a soil survey, each farmer should choose the best methods of conservation and soil and water management for his farm.

Wastelands should be stabilized with permanent, undisturbed vegetation to prevent damage to more valuable farm areas. Tree farming and permanent pasture are profit-producing uses for farm areas unsuitable for cultivation. Both trees and permanent pastures need care if they are to produce large crops. Grasses and other close-growing plants are valuable when used in strips, waterways, headlands, and rotations, and for green manure and cover crops.

Terraces are conservation and management devices used on lands too steep to permit strip cropping, but not so steep that permanent pasture or woodland must be used. Another kind of terrace is used to hold water on dry, sandy flatlands.

Strip cropping is a variation of grassland farming, by which strips of sod or other close-growing crops are alternated with strips of cultivated crops. Crop rotation and contour planting are often used with strips. Types of strip cropping are contour strip cropping, field strip cropping, wind strip cropping, and buffer strip cropping. Contour strips may be laid out with both edges of the strips on the contour lines, with sets of even-width strips measured from a "key" or "base" contour line, and with alternate even- and irregular-width strips. Width of strips varies with local conditions.

Crop rotation is the growing of a selected number of different kinds of crops in regular order on any particular field. Rotation serves to maintain crop yields; conserve soil and water; improve tilth; maintain organic matter; permit more profitable use of fertilizers, lime, and manure; keep the soil occupied; change the feeding range; counteract toxic substances; and improve crop quality. Many factors determine the crop rotation. Erosion hazards may be too great to permit successful rotation.

Farm ponds are another valuable device for soil and water conservation and better farm management.

Study Questions

1. What factors determine the kind of crop rotation to use?
2. What is to be gained by a crop rotation?
3. Why should farmers plant both shallow- and deep-rooted crops in rotation?
4. When should manure and fertilizer be used in the rotation?
5. What determines the selection of crops for a rotation?
6. What has been learned from the Morrow Demonstration Plots?
7. What is to be gained from the use of green manures and cover crops?
8. What kinds of strip cropping are used?
9. What practices are used in laying out contour strips?
10. What practices are used in maintaining terraces?
11. What are the values of grass waterways and grass headlands?
12. What special tillage practices are observed in strip farming?

13. What are the values and practices of grassland farming?
14. What is a good size and location for a farm pond?
15. How effective is tree planting as a conservation practice?
16. Why should experts make a pond survey?
17. What are the recommended practices in pond construction and maintenance?

Class Activities

1. Determine the crop rotation practices in the community.
2. Study effects of continuous cropping practices on yields and soil erosion.
3. Learn the community practices regarding strip cropping and use of grass waterways.
4. To what extent is grassland farming being used in the community?
5. Cite local examples of use of tree plantings to control erosion.
6. Make a field trip to a farm pond.

Conservation of Water and Soil in the Cultivated Field

Losses of water and soil are at their highest during periods of cultivation. In the experiment conducted by the Missouri Agricultural Experiment Station, you remember, the average annual soil loss per acre from bare, cultivated, uncropped land was 41 tons. Average annual rainfall loss through runoff was 30 per cent. During the same experiment, well-covered soil annually lost only 0.3 tons of soil per acre, and only 12 per cent of rainfall through runoff.

The farmer's soil and water resources are never in greater danger, then, than during periods of planting, cultivation, harvest, and winter or summer fallowing when bare soil is exposed. During such periods the farmer is running the risk of losing the fine particles that give topsoil its greatest value. He is running the risk of losing water that might be used by crops. He is running the risk of losing valuable organic matter, manure, lime, and fertilizer. He is running the risk of having his fields cut by rills and ruined by gullies. Lands below exposed fields may be overlaid with infertile subsoil washed downslope. The farmer's buildings may be injured by floods and his machinery by wind-blown grit. It should be noted, too, that all of these dangers decrease only slightly even during full vegetative maturity of such open-growing crops as corn.

Many of the measures treated in the preceding chapter will help control water and soil losses from cultivated areas. Stabilization of wastelands, tree farming, and grassland farming all serve in varying degrees to cut the amount of water running onto cultivated lands during storms, to hold water and snow on the land and add water to the soil, to lessen the force and angle of raindrops, and to lessen the force of wind. Terraces conduct runoff from the field slowly enough to help prevent rills and gullies, permit eroded soil to settle, allow some rainwater to penetrate the ground, and serve to an extent to slow wind speed on the ground surface. Grass strips help control wind and catch soil and water running off cultivated areas. Rotations have similar effects. Green manures which have been plowed under will help hold soil with intermixed roots and topgrowth.

Within the cultivated areas themselves, however, additional water and soil conservation measures are needed. These will be discussed below.

Water Conservation in the Cultivated Field

In Chapter 5, "Soil Moisture," you learned many of the devices for conserving water. All plants, you remember, need large amounts of water for growth. Water is not only an essential plant

Figure 19-1: Increasing Water Entry. The board at right is covered with a blotter, that at left with slick cardboard. Like a blotter, soil with a good water entry rate has a relatively rough, loose surface, which may be supplied by plowing. A cover of vegetation or crop residue has the effect of adding a thick layer of shredded blotter on top of the first blotter. You can demonstrate the value of contour plowing by folding the slick cardboard into a series of horizontal ridges and stopping the ends with waterproof tape. Water will soak into the most compact soil if held on the surface rather than allowed to run off.

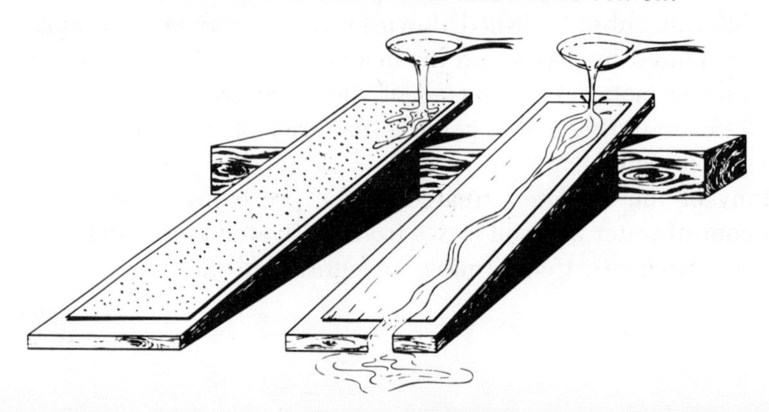

nutrient, but a solvent by means of which many other nutrient materials are made available to plants. Water is essential to soil microorganisms. The problems of conserving water in the cultivated field are (1) increasing water entry into the soil, (2) improving soil water movement, and (3) increasing water storage capacity.

Increasing Water Entry. Plowing is the most common device for increasing water entry. Plowing will break the crust that often seals the surface of cultivated soil, and will create a number of artificial pores or air spaces which permit more rapid water entry. Since plowed soil will settle fairly rapidly, these effects are temporary.

Contour plowing—plowing in level lines across a slope—is an extremely valuable water conservation device. Contour plowing serves to create a number of tiny dams, holding water on the soil and permitting it to penetrate.

Stubble and trash mulches also serve to break the crust-forming force of raindrops and to keep rainwater clear so that it will penetrate soil more readily. Mulches help hold water on the soil much as do plant stems and leaves. Sod-seeding is an allied practice of considerable benefit. Trash farming and sod-seeding are among the most effective devices for improving water entry rate, but they have the disadvantage of requiring special equipment.

Early planting and contour planting are other devices for increasing entry rate of water. Early planting works well only with close-growing crops, however.

Cover crops planted after harvest of cash or feed crops will serve to prevent runoff and to hold snow. Green manure crops which have been plowed under will help by increasing the number of soil pores.

All of the above practices that involve ground cover are most valuable in relatively humid regions. In dry regions, ground cover is often a disadvantage because it serves to intercept moisture which evaporates without reaching the soil. For these regions, summer fallow is recommended to increase water entry rate. Wind protection is usually necessary to prevent soil erosion from lands under summer fallow.

Improving Soil Water Movement. Soil water movement, you remember, is mainly controlled by such factors as soil texture, aeration, and content of organic matter. Little can be done to improve the movement of soil water in lands actually under cultivation and growing a crop. Plowing does have a temporary effect, but in the plow layer only. In general, then, provision must be made for improving soil water movement before or after cropping. This is usually accomplished by plowing in animal or green manures, and by plowing in crop residues. Grasses grown in rotations have a good effect on soil water movement.

Increasing Water Storage. The water-storage capacity of the soil is determined by the size of soil particles and the organic content of the soil. As with water movement, storage capacity of soil can only be improved before or after cropping, with the use of green manures, animal manures, and crop residues plowed into the soil. Both crop selection and control of evaporation, however, will diminish the amount of water used by plants or lost from the surface. The devices for increasing water entry, such as contour plowing and summer fallowing, will increase the amount of water a cultivated field will store. Increased entry rate during cultivation is especially important with clay soils, which tend to admit water slowly but store it very well. Devices for diminishing the amount of water evaporated from the surface of cultivated soil include summer fallow, mulches, and weed control.

Soil Conservation in the Cultivated Field

Along with overgrazing and unwise cutting of forests, cultivation is what makes soil erosion possible. On lands well covered with vegetation, little soil erosion takes place. There are two kinds of soil erosion, *water erosion* and *wind erosion.*

Water Erosion

Common types of water erosion are sheet erosion, rill erosion, gully erosion, and bank erosion. An undesirable result of these forms of erosion is soil deposition. Soil is eroded by water in two stages, (1) detachment, and (2) transportation. Water erosion never becomes severe unless both detachment and transportation take place. A knowledge of what all these terms mean and how the forces of water erosion work assists in developing conservation practices.

Kinds of Water Erosion. As stated above, common types of water erosion are sheet, rill, gully, and bank erosion.

Sheet Erosion. In sheet erosion, a fairly uniform layer of soil is removed from an entire surface area. This is caused by splash from raindrops. Much of the soil loosened by splash is actually transported from the field in rills or gullies.

Rill Erosion. Rill erosion is caused by water running in very small streams over the surface of the soil. Small channels are cut by the running water itself, and by the soil particles it contains. These channels are small enough so that subsequent tillage fills them. The loss consists mostly of surface soil.

Gully Erosion. Gully erosion occurs when rills flow together into larger streams. These larger streams cut channels that ordinary tillage does not remove. Thus, they tend to remain in the field and to be cut deeper with each successive rain. Gullies will cut a field into small units and ruin it unless control measures are taken. Bridges may be required to cross them with farm machinery or livestock. They can be started even by wheelmarks.

Figure 19-2: Rill Erosion. Though rills can be filled by ordinary tillage, they are costly in terms of topsoil and the expensive fertilizers that may have been added to it. Note the evidence of soil deposition in the foreground.

They tend to stabilize against further cutting only as their bottoms become level with their outlets.

Bank Erosion. As the name suggests, bank erosion is the cutting of banks by streams and rivers. It is usually most severe on the outer sides of bends. It is very serious in time of flood, when it frequently destroys land, buildings, roads, and other valuable property.

Soil Deposition. Soil deposition is a *result* of the above forms of erosion. When the discharge water from a gully reaches relatively level land, the rate of flow decreases. Much of the soil material that was carried by the faster-flowing water is deposited. This sediment, which may be infertile, often covers crops to depths ranging to several feet.

Erosion by Soil Detachment. The first of the two stages of water erosion is soil detachment. Soil particles on fields either bare or planted with open-growing crops such as corn are held in place by roots and root fragments, gravity, the filaments of microorganisms, the viscosity (stickiness) of clay particles and particles of organic matter, electrical attractions, the binding action of irregularities of particles and soil aggregates, and similar forces. Before a soil particle can be moved from a location, all these forces must be broken down. In water erosion this is accomplished by either (1) the beating action of raindrops, or (2) the force of running water.

Soil Detachment by Raindrops. The effect of raindrops is known as *splash erosion.* Falling raindrops act like small bombs. Each time they hit the soil, they dig little craters and throw debris. The total energy of raindrops during a rain of one tenth of an inch per hour is estimated at 100 horsepower. Two inches of rain falling in an hour are estimated to provide force sufficient to lift a seven-inch layer of topsoil 86 times to a height of three feet.

The erosive capacity of raindrops varies with the diameter and the velocity of the drops. Large drops during a hard thundershower may have many times more force than those in a fine drizzle. Raindrops vary in size from a fine mist to nearly eight millimeters, or one-third inch, in diameter. The speed of falling raindrops increases with their size and with wind velocity. Even in a normal rainfall, soil particles may be thrown to a height of

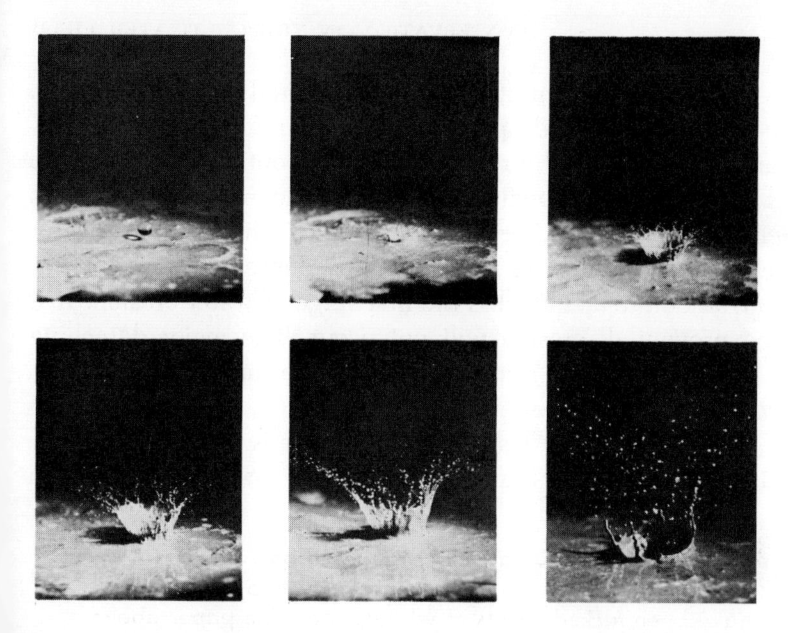

Figure 19-3: Falling Raindrop. The magic of the high-speed camera here captures the bomb-like impact of a single raindrop on muddy water.

two or three feet and to a horizontal distance of five feet. Larger soil particles, however, may be moved only a few inches or not at all.

At the same time, detached particles of soil that are not eroded from a field may have had their structures broken down. Thus they lie close together and form a relatively water-impervious layer over the surface. Such a layer decreases water infiltration and consequently increases runoff.

Raindrops are active in detachment in another way when they cause turbulence in shallow pools of water. This turbulence, or *whirlpool action,* can be observed during a heavy shower.

Soil Detachment by Running Water. The second detaching force is running water. The speed at which water flows, which is controlled by slope and volume, largely determines its detaching force. It should be noted also that, during a storm, water tends to form into small ponds on fairly level land and to flow quite rapidly between ponds when they differ slightly in level.

The effectiveness of moving water in the detachment process is greatly increased by the presence of soil particles in the water.

These bombard adjacent attached particles and carry them away.

Detachment by raindrops occurs fairly uniformly over an entire area. Detachment by running water is confined only to those areas where water is flowing. Raindrop detachment thus tends to loosen soil particles uniformly over a field, while running water detachment is confined to rills, gullies, and streams.

Control. Soil detachment is the first stage of water erosion. If it can be prevented, erosion will not occur. The major detaching force is the beating of raindrops. When the impact of raindrops is decreased, detachment is also decreased.

As you have already learned, plant cover is a highly effective means of reducing the impact of raindrops. With many plants, a

Figure 19-4: Controlling Detachment by Raindrops. You can repeat this simple experiment with a hose and two boards upon which have been tacked sheets of white paper. The paper above the sod of your lawn will stay clean; the paper above bare soil will become dirty. Early planting, fall planting, intercropping, cover crops, and mulches help protect the soil of cultivated fields.

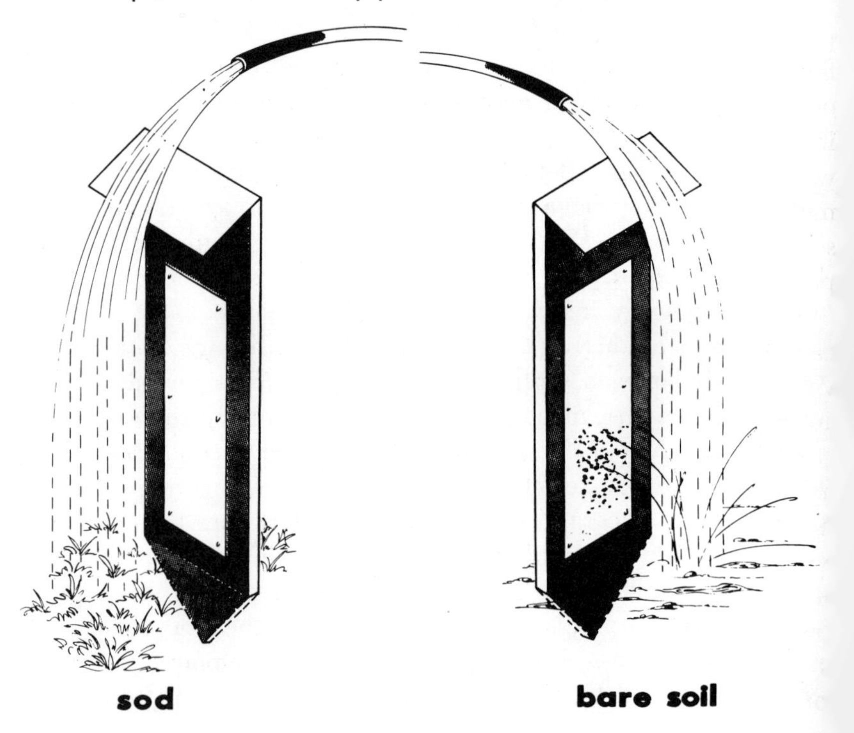

high proportion of the intercepted rainfall flows down the plant stem at a slow pace. When stems are close together, they act as anchors for small dams built up by the organic and other debris which is almost universally present in a dense stand of plants. Thus, dense growth will retain much water on the soil surface and only a limited flow will take place. Some vegetation, such as grass, tends to "lodge" under a heavy flow of water. This forms a thatch over which water flows without much opportunity for detaching soil. Finally, plants tend to keep water spread uniformly over the land surface, helping to prevent rill and gully erosion.

In annual cropping, then, an effort should be made to provide a good plant cover during the stormy season when the soil detachment hazard is greatest. Early planting and fall planting are highly valuable, and so are cover crops grown during the parts of the season that major crops are not grown. Anything that helps regular crops grow well—such as fertilization or irrigation—also increases ground cover. Some open crops such as corn are *interplanted* with rows of legumes.

Crop residues and mulches of hay, straw, and wood chips have proved to be an excellent substitute for plant cover in cultivated fields. Though special equipment such as the mulch planter is often required for successful "trashy farming," results seem to indicate that soil erosion is held to a minimum and that good water conservation is also provided. Manure may be added to mulches to help hold them in place. Other farmers protect their soils by leaving standing stubble in the fall and during fallow periods.

Though the use of green manure does not strictly apply to the care of soil while it is actually growing a cultivated crop, cultivated soils to which a green manure crop previously has been added do tend to resist detachment.

All of the above soil conservation practices are also effective, in varying degrees, at reducing detachment by running water. Additional aids within the cultivated area are plowing, which will destroy rills, and contour plowing, which will create a series of small dams across a field. Small gullies should be filled with a grader or some similar equipment. Sometimes a slight depression is left and former gullies are converted into grassed waterways for drainage purposes. Mulches and other mechanical aids may

Figure 19-5: Transportation by Raindrops. Raindrops have lifted a deep layer of soil from this field and transported it far away. Even flat fields can be severely eroded in this way.

be needed to get grass started. Large gullies must be stabilized with permanent vegetation. Trees and grasses are recommended for this purpose. Dams of various sizes are often used also. Cultivated fields which show considerable erosion damage are probably being improperly cropped. Such mechanical aids as terraces may be required, or vegetative treatment such as use for permanent, carefully grazed pasture.

Soil detachment from the banks of streams and rivers is sometimes controlled with bulkheads of brush or rock. Willows and similar water-loving trees and plants will provide considerable protection also.

Erosion from Soil Transportation. After detachment has taken place, soil particles are also transported by the action of both raindrops and running water.

Heavy showers are frequently associated with heavy wind. The raindrops then hit the ground at an angle rather than falling straight. The splash thus tends to move downwind. While the splash from each drop of water moves soil only a few feet, the total force of a heavy rain with high winds will move large amounts of soil over a considerable distance by repeated splashes of soil particles downwind. It is easily observed that mud is splashed to a height of three feet or more on a building surrounded by unprotected soil.

Much of the soil detached by raindrops is also carried long distances by running water. The ability of water to transport soil particles increases tremendously with the speed of its flow.

An examination of a bare field following a heavy shower provides information on the nature of the erosion pattern. Since splash erosion is quite uniform over the field, you have to look carefully to see evidence of it. If you do look carefully, under a leaf or a pebble or anything else that has protected the soil from the force of the raindrops, you will find that the protected soil is slightly higher than that on the unprotected surrounding area. The rills and gullies cut by running water, on the other hand, can be seen readily. There are also miniature deltas where transported soil has been deposited as the flow of the water has decreased.

It can frequently be observed that the soil deposited in the deltas is more sandy than the soil which was not moved. This is because fine particles are less likely to settle out than coarse particles. Erosion tends selectively to remove the fine particles containing the major portion of the nutrient elements, and the surface soil containing most of the organic matter.

Control. Soil transportation may be partially controlled within cultivated areas by conservation measures taken outside them. The force and angle of falling raindrops may be limited with windbreaks and to some extent by strip cropping. Transportation by running water may be limited with shelter belts, strips, and similar devices.

Control measures on the field itself are similar to those for soil detachment. Plant cover, residue or mulch cover, contour plowing, and rill and gully control are all effective in varying degrees. Field drainage terraces are also helpful.

Wind Erosion

To understand wind erosion, we must first know something about the pattern of wind movement. When the wind blows, we can feel that the air is not moving uniformly. We often say that the wind is *gusty,* meaning that the wind velocity is highly variable. This can readily be seen by watching the movement of smoke. By watching smoke, too, you can see that air currents move up and down as well as parallel to the surface.

The surface of the ground offers resistance to the flow of air over it. Thus velocity is much slower close to the ground surface than it is at higher elevations. Anything that increases resistance to air flow decreases wind speed. The slower we can keep wind movement adjacent to the soil surface, the less chance there is for wind erosion to take place.

Fields usually have a few spots where wind erosion is most likely to start. These may be areas more exposed to the wind than others, or they may be areas containing soil particles of a size especially susceptible to wind erosion. Such particles are extremely fine. Larger particles are too heavy to move readily, while particles still smaller are too close to the general level of the ground.

How Wind Erosion Starts. When wind speed reaches the *critical level,* particles start to move by rolling along the surface. After particles have rolled a short distance, they tend to bounce into the air. Here wind movement is faster. Thus, at the end of each bounce the particles are traveling downwind at a faster rate than at the start of the bounce. When the soil particles again hit the surface, they either bounce once more or start other soil

Figure 19-6: Windbreaks. The principle for controlling wind erosion by measures taken within the field is the same as that for controlling wind erosion by measures taken outside it—creating a barrier that keeps wind above the surface of the soil.

Figure 19-7: Free Flight. Spectacular dust storms must be prevented by conservation measures at the soil surface.

particles to roll and bounce. The bouncing action of soil particles continues throughout a windstorm, and is responsible for more than half of the damage caused by wind erosion. Most of the soil particles traveling in this way are one foot or less from the surface.

In addition to bouncing, there are two other ways in which particles are moved by wind. One of these is known as *free flight.* Particles traveling in this manner are small enough to be carried long distances and hence form the spectacular dust storms. Since these small particles are usually kicked from the soil surface by larger bouncing particles, their number never becomes large until after bouncing has started.

The third type of soil movement caused by wind is known as *surface creep.* Soil particles too large to be rolled are moved downwind by constant bombardment from smaller bouncing particles. Even pebbles and small stones may be moved in this way.

Control. Since the bouncing of soil particles by wind causes all of the types of wind erosion, control of bouncing also will control free flight and surface creep. There are three important ways in which control can be accomplished in the cultivated field. All involve slowing the wind movement at the soil surface.

The first method involves the use of plants or trash on the soil surface.

Figure 19-8: Grass Bands on the Contour. You can see what a good job of controlling wind erosion these grass bands are doing. Contour planting makes them helpful in conserving rainwater and snow as well.

The second method is accomplished by plowing. The purpose here is to form soil into clods or aggregates too large to be readily moved by wind. Because soil clods tend to break down even during light rains, the effectiveness of the cloddy surface is usually temporary.

A third method is to prepare the soil surface so that the first particles that start to bounce are trapped after they move a short distance. A ridged surface tends to trap these particles in the furrows. To do this successfully, the ridges must contain some material coarse enough to stabilize them, so that the ridges themselves do not lose soil particles.

Important protection against wind erosion may also be given by soil management outside the cultivated areas. In dry areas, wind strip cropping has proved effective. Windbreaks are also helpful. In both cases, particles that start to move in the cultivated area are trapped by the strip crop or the trees.

Summary

Losses of water and soil are at their highest during periods of cultivation. Conservation measures such as stabilization of wasteland and grassland farming in areas adjacent to cultivated areas help give cultivated lands protection. Within the cultivated lands themselves, additional water and soil conservation measures are needed.

Water may be conserved by increasing water entry into the soil, improving soil water movement, and increasing water storage capacity. Water entry rate may be improved by plowing, contour plowing, mulching, prompt planting, and using cover crops. In dry regions, summer fallowing is used. Little can be done to improve water movement in lands under cultivation and growing a crop. The amount of water stored by a soil, though not its storage capacity, may be increased during cultivation by increasing entry rate and decreasing surface evaporation.

Cultivated soil is eroded by wind and water. Common types of water erosion are sheet erosion, rill erosion, gully erosion, and bank erosion. Soil is eroded by water in two stages, detachment and transportation. Detachment is accomplished by raindrops and running water. Control measures include plant and trash covers, plowing to destroy rills, contour plowing to dam runoff, and filling or checking and stabilizing gullies. Bank erosion may be controlled with rock or brush bulkheads or willows. Soil transportation, following detachment, is also accomplished by raindrops and running water. Control measures are similar to those for detachment.

All types of wind erosion are caused by bouncing soil particles. Wind erosion is controlled by slowing wind movement at the surface of the soil. This is accomplished with plant and trash cover, with plowing the soil into clods, and by ridging the soil surface to trap bouncing particles. Measures outside the cultivated area, such as windbreaks and wind strip cropping, are also effective.

Study Questions

1. Why is erosion particularly dangerous on bare fields?
2. What are the two major processes involved in water erosion of soil?
3. What is the major detaching force in water erosion?
4. What is the major transporting force in water erosion?
5. How does splash from raindrops transport soil?
6. Name and describe the various kinds of water erosion.

7. How does erosion damage the soil remaining in the field?
8. What are the problems of water conservation?
9. What are the three ways in which soil particles are moved by wind?
10. Why is it essential to stop the bouncing of soil particles to control wind erosion?

Class Activities

1. Plan a trip to watch effects of water and wind erosion during a rain.
2. Use an ordinary garden hose to compare the effects of a full stream of water on grass sod and open soil.
3. Observe throughout the community erosion-made formations such as hills, gullies, and valleys.
4. Determine the amount of soil removed by the force of natural or artificial rain from a flat box containing uniform soil material.
5. Demonstrate how rain puddles, compacts, and seals the soil surface.
6. Demonstrate how raindrops cause sand particles to move downhill on sloping land.

Tillage Practices and Equipment

20

Since cultivation or tillage is perhaps the major cause of soil erosion, it is essential that farmers understand the exact nature of its objectives and its limitations. Many farmers have permanently ruined their fields with tillage practices that could not supply the benefits expected of them. Other farmers have wasted large sums of money on unnecessary or even harmful tillage practices. Tillage is one of the most costly items in crop production. Any tillage practice that does not return its cost in increased production or soil maintenance should be eliminated or altered. When a cheaper operation can be used satisfactorily, profits are increased.

The objectives of tillage are temporary soil conditioning, seedbed preparation, water conservation, aeration, weed control, and utilization and disposal of crop residues.

The long history of tillage, which has been practiced since crops were first domesticated, may be the principal reason that these objectives are sometimes mistaken or misunderstood. When a relatively new farming technique is introduced, farmers and soil scientists are careful to study it. With experiments, they determine what it can and can not do. Tillage, on the other hand, too often is taken for granted. During the last 100 years, there has been a tremendous development of tillage equipment. This development, however, has been made less from the standpoint of crop needs than from the standpoint of engineering. In many cases, traditional tillage practices themselves have been assumed to be satisfactory. Engineers have devised machinery which will perform traditional tillage practices more rapidly and with less physical labor. If the equipment worked and crops grew as well as they had before, the equipment was thought all right. Tillage equipment, in other words, has saved the farmer time and work, but it has not necessarily improved yield or quality of crops.

While we do know fairly well what the objectives of tillage are, then, in many cases we must perform them with machinery which has not been specifically designed for the task. Part of the farmer's job is to choose the equipment that will come closest to meeting one or more of the desired objectives.

We also are under the handicap of not knowing exactly how the objectives of tillage may best be met. For instance, we know only in general terms what a good seedbed is. Similarly, the amount of aeration necessary for optimum growth of different crops is not known. Once soil scientists learn these things, it may be possible for our engineers to devise better tillage equipment than we now have.

Meeting the Objectives of Tillage

Temporary Soil Conditioning. Tillage leaves the soil temporarily looser than it was. It increases the size of soil pores and thus aids air and water penetration. It makes the soil more easily penetrated by roots.

Figure 20-1: Temporary Soil Conditioning. Loosening the soil with tillage permits air, water, and roots to penetrate more easily. These benefits are usually confined to the plow layer, however, and are lost as soon as natural conditions such as rainfall make the soil revert to its original state.

At the same time, tillage should not be expected to act as a substitute for permanent soil improvement. The major elements determining the physical condition of the soil are texture and structure. These are controlled by the minerals and organic matter which make up the soil, and by the climate, particularly the rainfall, which has a noticeable influence on the arrangement of minerals and organic matter. Together, these factors largely determine soil tilth or workability. Tillage cannot have more than minor effects. The forces that tended to arrange the soil particles in the first place are still active, and after a few cycles of wetting and drying or freezing and thawing the particles and aggregates will have returned to their original arrangement. Many farmers recognize this fact and yet repeat the same, unsatisfactory tillage operations year after year.

Seedbed Preparation. For a seed to germinate, it must have (1) adequate moisture, (2) a satisfactory temperature, and (3) a certain amount of oxygen. After a seed germinates, its roots must be able to grow through the soil to secure nutrients and water, and its top growth must be able to break through the crust on the soil surface.

To provide moisture for germination, the seed must have a good contact with moist earth. The soil clods and aggregates must be fine enough so that the seed can be closely surrounded. The smaller the seed, the more important it is to have a fine seedbed. Small seeds must also be planted close to the surface to allow the seedlings to emerge. It is important to note in this connection that a shallow layer of soil above a seed dries quickly. Fine seeds must be able to germinate in time to push roots to greater depth where the moisture supply is more constant. Larger seeds make better contact because of their size. In addition, they are able to sprout through a thicker layer of soil and thus can be planted at greater depth, where the soil will not dry out.

Climate is the controlling factor of soil temperature. Seeding dates must be adjusted to insure sufficient warmth. Excessively wet soils are slow to warm in the spring, and seeding and germination are frequently delayed. Good soil drainage helps prevent this.

If moisture conditions are satisfactory, there is usually enough air around the seed for germination. Air supply for downward

Figure 20-2: The Value of a Good Seedbed. The fine seedbed at right has made nutrients and moisture more readily available. Earlier sprouting conserves soil and water by providing earlier ground cover. Weeds will make poorer growth because they will be shaded by the crop. The crop that sprouts first usually reaches maturity first, and often brings a better price.

root growth may become critical, however, even when the surface layers have an adequate supply of oxygen. If soil is too dense, roots also have trouble penetrating it even though they exert fairly high pressures in growth. Tillage loosens soil so that roots can penetrate more readily. As the season progresses, however, the effects of tillage are reduced because the soil compacts from a series of wetting and drying cycles.

It is usually impractical to till the soil to a depth of more than a few inches each year to influence the rootbed, but in some areas an occasional deeper tillage or *subsoiling* can be justified. While a satisfactory rootbed is important to plant growth, it is difficult to attain through cultivation.

Planting. Planting as a tillage operation is receiving more and more attention. An example of the practice is the planting of corn in a single operation with a planter that immediately follows the plow. Planting on a contour permits timely planting of fields since bottomlands are often still too wet for planting when upland soils are dry. Planting of grass with small grains using combination drills is a common practice. Planting of wide rows of corn with legumes or grasses between rows is a two-crop system of soil management. The legume crop may be plowed under as green manure for the following year, or left as a perennial crop.

Depth of Planting. Depth of planting is an important consideration. Small seeds such as those of clover and timothy are planted one-half inch to one inch deep; small grains, one to two inches deep; corn, 1½ to 2½ inches deep; and potatoes, three to four inches deep.

Cultivating Row Crops. Cultivation of row crops conserves moisture, aids aeration, and provides weed control. The farmer should keep in mind, however, that the root growth of young plants is often more extensive than top growth. Considerable damage may be caused by deep cultivation. Roots of corn, for instance, grow near the surface and meet between rows by the time the plants are knee-high. Close cultivation is needed for weed control. Cultivating wet soils will disturb their structure and produce hard clods.

How many times to cultivate is a much-debated issue. In general, cultivation should only be frequent enough to control weeds and to provide for moisture and air penetration by breaking the surface crust of the soil.

Water Intake. Surface tillage before rains will break the soil into lumps with relatively large pores between them. This allows

Figure 20-3: Cultivation in Strips. To help control wind erosion, this corn is being cultivated in strips.

water to enter the soil readily, because absorption depends upon the size of the openings between particles. Such tillage is especially valuable with the less-absorbent soils, permitting them to store water in the surface irregularities and the large pores left by plowing. This water then has time to enter the slowly permeable subsoil.

Tillage is less necessary in open, coarse-textured, sandy soils, which generally absorb water readily even when they have some clay. Other, fine-textured soils are so well aggregated that they take water about as readily as sands. Soils of good tilth have stable clusters of aggregates within which pore spaces are small but between which the openings are relatively large.

It should be noted both that tillage affects the pore size only in the tilled layer, and that beating rain will break down surface soil aggregates. Fine soil particles are carried into the openings with the water, clogging them and thereby "waterproofing" the soil surface. The farmer thus also must use such moisture conservation practices as terraces, strip crops, and contour plowing.

The value of a dust mulch as a water conservation device is debated. Many soil scientists believe that about as much water is lost in establishing the mulch as is saved in subsequent evaporation losses.

Air Movement. Plant roots take oxygen from the soil air and release carbon dioxide. In an open soil, soil air is constantly being interchanged with the air of the atmosphere by a process called *diffusion.* To make diffusion possible, there must be continuous pores through the different soil layers to the surface Tillage increases the proportion of large pores in the tilled laye and thus increases the supply of oxygen. It also serves to breal the crust that forms on the soil surface and permits more air t enter the soil. The farmer can tell when such tillage is necessar by watching for yellow, sickly leaves. The removal of exces water by drainage is the most practical way of improving soi aeration at depths greater than that of cultivation.

Weed Control. Our cultivated crops have been pampered b man for centuries. They cannot compete with weeds for nutri ents, light, and water. Cultivation controls weeds either by (1 cutting their roots, (2) separating their roots from the soil, ((3) covering the entire plant.

Figure 20-4: Weeding Corn. Note that this tillage operation is also breaking the crust that has formed on the soil. What will be the result?

During the tillage operation, the roots of weeds can be severed from the tops by the tillage implement. Weeds must be cut off deep enough so that the crown of the plant is severed from the roots, but if tillage is too deep enough roots remain active above the point of shear to maintain the plant until it can establish new roots. Plants with deep roots must be kept sheared off near the ground surface so that eventually the roots are starved to death.

For control of plants with shallow roots, tillage can move the roots to the surface, where they are killed by drying. Weed seedlings also may be killed by the method of separating roots from the soil. It is essential that the soil undergo considerable manipulation during the process, but shallow cultivation is all that is required and this can be done fairly economically. Another advantage of shallow tillage is that it does not move buried weed seeds to the surface. Since many weeds do not germinate when buried at considerable depth, it is a weed control measure to leave them there.

Covering weeds is another method of control, since covering a plant cuts off its source of light. Most plants must be completely covered, however, and this requires deep, thorough tillage. The plow is the most effective covering implement.

Annual weeds are much easier to kill than perennial weeds. All three of the above methods are effective with annual weeds. These methods also work with perennial seedlings. Established perennial weeds have root stalks that send up buds after tops have been cut. Thus, the roots must be killed. This requires keeping top growth at a minimum by frequent cultivations, so that the plants never grow enough leaves to build up food reserves.

Weed control by chemicals is becoming popular. Chemicals can often be used to supplement or replace one or more weed-control tillings.

Handling Crop Residues. Crop residues are a source of organic matter needed for soil maintenance. By proper handling, crop residues can be made to serve additional purposes. Left on the surface, they prevent soil puddling by protecting the soil from raindrops. Large pores thus are kept open for rapid water infiltration and aeration. In addition, residues slow runoff and surface wind and thus decrease erosion. Leaving small-grain crop residues on the surface for wind erosion control by tilling with shovel- or blade-type implements is common on the Great Plains. Use of residue for protection against water erosion is also becoming common in the production of row crops.

Heavy residues frequently cause difficulty in seeding and tillage. Special equipment is available which will do a good job of seeding in spite of moderate amounts of residues. It is possible to leave heavier residues in humid areas, where such materials break down rapidly when in contact with the soil.

Figure 20-5: Chemical Weeding of Rice. Tillage is not the only method of controlling weeds, which compete with crops for light, nutrients, air, and water. The pictures contrast one portion of a rice field sprayed with a new chemical and another part of the same field which was not sprayed.

Figure 20-6: Moldboard Plow. The moldboard plow is among the commonest of tillage implements. It is used for temporary soil conditioning, preliminary preparation of the seedbed, and turning under crop residues and green manures.

Tillage Equipment and Its Functions

There is a variety of tillage equipment designed to perform such specific tasks as loosening soil which is too compact, packing soil which is too loose, breaking up soil clods or crusts, controlling weeds, and covering or breaking up trash.

Major categories of tillage equipment include plows, disks, harrows, blade or shovel tillers, and subsurface tillers. A number of other types are included in the following discussion.

Moldboard Plows. The moldboard plow is one of the most widely used tillage implements. It is excellent for loosening firm soil and breaking up tough sod. It will turn under heavy amounts of green manure, crop residues, and other trash. Under ideal conditions, it also tends to encourage aggregation. Newly plowed land presents a rough surface which provides large pores for ready penetration of water. The furrows also provide for temporary ponding during rains.

The plow is designed to cut, lift, and invert the furrow slice. A short moldboard plow turns the soil quickly and provides maximum compression and shattering forces. A longer moldboard more completely inverts the furrow slice without shattering, and is adapted to turning sod.

In some northern areas there is a tendency to plow heavy soils in the fall so that freezing and thawing during the winter will break up clods. However, fall plowing buries crop residues which may be needed to hold snow and to prevent erosion and surface puddling. Local conditions must be considered in deciding between fall and spring plowing.

Depth of plowing is an important problem. On open, permeable soils, shallow plowing is frequently satisfactory. Deeper plowing will incorporate more residue from a previous crop when this is desirable. On compact soils, deeper plowing opens the soil to greater depth. Because consumption of tractor power increases with increased depth of plowing, it is desirable to plow as shallowly as possible, while meeting the objectives of the operation.

Disk Plows. Many disk implements are available for tillage operations. These differ in the size of the disks, the angle at which the disks run through the soil, and the amount of curvature of the disks. One common type of disk implement is the disk plow.

Disk plows work well in soils too dry or hard for a moldboard plow to penetrate. They work better than a moldboard plow in sticky soils such as gumbo and hardpan, since moldboard plows

Figure 20-7: Disk Plow. The disk plow shares the uses of the moldboard plow, but is superior for certain soils. After studying the text, name these soils.

Figure 20-8: Spike-Tooth Harrow. The spike-tooth harrow is here being used to finish a seedbed.

scour poorly in such soils. Disk plows work well in loose ground and in stony land. Sharp disk plows can cut and turn furrow slices. Heavy-duty disk plows can penetrate as deep as 20 inches in some soils. Two-way disk plows are available.

Disk Tillers or One-Ways. Disk tillers have many features in common with disk plows, but have closer disks which are set to throw the soil in one direction. As a result, the furrows are smaller. Disk tillers are used extensively for preparing wheat land in the dry areas of the Western states. In continuous cropping operations, seeding frequently follows tillage.

Disk Harrows. The function of disk harrows is to pulverize and pack the soil, leaving a surface mulch and a compact subsurface. Disk harrows used before plowing serve to break the surface and mix the trash with the topsoil. Disk harrows used after plowing serve to pulverize lumps and close air spaces in the turned furrows. Disk harrows are made as single action and double action. The single action type is used to rough grain stubble and work summer fallow land. The double action type is used for general seedbed preparation.

Rotary Tillers. Some authorities on seedbed preparation suggest the use of rotary tillage equipment. This prepares a finely pulverized soil. Such a seedbed tends to break down the natural soil aggregates, however.

Spring-Tooth Harrows. Spring-tooth harrows dig three to four inches into the surface soil. They are effective in cultivating stony fields and in pulling weeds.

Spike-Tooth Harrows. Spike-tooth harrows are effective in leveling and stirring up fairly even fields. They are of use in killing weeds both prior to their emergence and immediately after.

Cultipackers. As the name suggests, cultipackers cultivate and pack down the seedbed. They are desirable in preparing a firm seedbed for small seeds such as those of clover.

Rod Weeders. Rod weeders are used extensively as a summer fallow tool. The weeder rods revolve slowly just below the ground surface in a direction opposite to the forward movement of the machine. They tear out weeds by the roots and bring them to the surface.

Blade or Shovel Tillers. Blade or shovel tillers are a broad group of implements. They consist of blades carried on shanks, which cut through the soil at a fairly constant depth. Two blades usually form a V, and the carrying shank is welded to the top. On some implements the base of the V is several feet wide, while on others it may be only a few inches wide. The blades are usually set so that they have some lift. This lift pulverizes the soil without inverting it. These implements are effective in cutting

Figure 20-9: Panbreaker. Panbreaking is part of the temporary soil conditioning that can be accomplished by tillage. Note how the point lifts and shatters a hard, dry plowpan. Fertile topsoil is kept on top.

weed roots from top growth, and are often used for weed control in row crops.

Blade implements do not turn the soil and thus sometimes are called *subsurface* implements. They leave all residues on the surface. They are used extensively in the Great Plains, where residue is left on the surface to discourage wind erosion.

Subsurface Tillers. Most tillage involves only the top few inches of soil. However, an occasional tillage to greater depth is beneficial to some soils. If the soil contains a compact, water-impervious layer which can be shattered or mixed with more desirable materials, the operation can favorably influence the soil for some time. When the subsoil is uniformly compact and impervious, however, the effect of deep tillage is likely to be insignificant. After a few wetting and drying cycles, the soil reverts to its original condition. Under these conditions, subsoil tillage, which requires very heavy equipment, is not likely to pay for itself.

Summary

It is essential that farmers understand the objectives and limitations of tillage because tillage is a major cause of soil erosion and because unnecessary tillage is a great expense.

The objectives of tillage are temporary soil conditioning, seedbed preparation, water conservation, aeration, weed control, and utilization and disposal of crop residues.

There is a variety of tillage equipment designed to perform such specific tasks as loosening soil, packing soil, breaking up soil, controlling weeds, and covering or breaking up trash. Major categories of tillage equipment include plows, disks, harrows, blade or shovel tillers, and subsurface tillers. The farmer must choose the equipment best suited to meeting one or more of the objectives of tillage.

Study Questions

1. List the major objectives of tillage.
2. How do tillage practices affect soil structure?
3. What is a good seedbed?

4. Under what conditions does water best enter the soil?
5. Under what conditions is the soil best supplied with air?
6. How effective is tillage for weed control?
7. What are the recommended tillage practices for handling crop residues?
8. What are approved practices in cultivating row crops?
9. What are approved practices in planting (1) small grains, (2) grass seeds, and (3) other crops?
10. Can farmers over-till their land?
11. What are the functions of (1) moldboard plows, (2) disk plows, (3) disk tillers, (4) disk harrows, (5) rotary tillers, and (6) subsurface tillers?

Class Activities

1. Determine the different kinds of tillage equipment used in the community.
2. Visit different kinds of tillage operations, such as (1) fall plowing, (2) summer fallowing, and (3) subsoiling.
3. Visit farm machinery salesrooms to study new kinds of tillage equipment and their uses.
4. Discuss soil tilth and effects of tillage operations with farmers.
5. Examine different tillage implements and decide which of the objectives of tillage they accomplish.
6. Plant grass and grain seeds at varying depth and note the percentages of plants that emerge.

Land Judging

21

Land judging, or *soil survey* information, is recommended for planning and coordinating soil improvement and management on each farm. Land judging gives the farmer a kind of inventory, which tells him exactly what he has to work with, and what improvements he needs to make. Included in this inventory are studies and ratings of surface soil texture, subsoil texture, depth of surface soil and subsoil, topsoil color, soil structure, stoniness, ease of cultivation, subsoil permeability (movement of air and water in the subsoil), surface drainage, internal or subsurface drainage, per cent of slope, degree of erosion, erosion hazard when cultivated, factors used to select land classes, and land capability classes. All soils of the farm are tested for fertility, reaction, and alkali conditions.

Each of these details is generally indicated by a symbol on a detailed map of the farm, which is divided into areas called *mapping units.* Mapping units are made up of soils having similar characteristics. Farmers then determine recommended conservation and management practices. Some mapping units will need vegetative handling, such as putting rocky slopes into permanent woodland. Others will need mechanical improvements, such as irrigation or terracing. Still others will need fertilizer, or soil amendments such as lime and gypsum.

Land judging can be learned in special schools, and many land judging contests are held. Since the score cards used in these contests provide a summary of the land judging practices

Figure 21-1: Edd Roberts at a National Land Judging Contest. Vocational agriculture students and instructors learn and practice land judging at local, state, and national contests.

that should be applied to each mapping unit, they are convenient devices for studying the considerations that make up the soil survey.

The first land judging school was conducted in 1941 at Red Plains, Oklahoma, by Edd Roberts, State Soil Conservation Extension Specialist, and Harley Daniel, Project Supervisor. The first national land judging contest was held in 1951 in the same state with 2000 youths from 15 states participating.

Most agricultural colleges have published bulletins dealing with land judging, and Edd Roberts is the author of a book on the subject published by Oklahoma University Press, Norman, Oklahoma. Today FFA and 4-H groups have local, district, and state land judging schools and contests.

Land Judging Score Cards

Most land judging score cards are uniform in describing the *internal* and *external* land features to be judged. Greater variation exists in recommended soil management practices, since these must be applied to local conditions. Listed below are some common score card items. For Part I, read down the left column first, then down the middle column, then the right column.

PART I

Physical Features of Land and Soil

Surface Soil Texture

Coarse
Medium
Fine

Subsoil Texture

Coarse
Medium
Fine

Depth of Surface Soil and Subsoil

Deep
Moderately deep
Shallow
Very shallow

Topsoil Color

Light
Medium
Dark

Soil Structure

Platy
Prismatic
Columnar
Blocky
Granular
Single-grain
Massive

Stoniness

Free
Slight
Moderate
Excessive

Ease of Cultivation

Not difficult
Difficult
Very difficult

Subsoil Permeability

Very slow
Slow
Moderate
Rapid

Surface Drainage

Poor
Fair
Good
Excessive

Internal Drainage

Excessive
Good
Fair
Poor

Per Cent of Slope

Nearly level
Gently sloping
Moderately sloping
Strongly sloping
Very steep

Erosion

None to slight
Moderate
Severe
Very severe

Erosion Hazard When Under Cultivation

None to slight
Moderate
Severe

Factors Used in Selecting Land Classes

Texture
Permeability
Depth
Slope
Erosion
Drainage

Land Class

Class I
Class II
Class III
Class IV
Class V
Class VI
Class VII
Class VIII

PART II

Recommended Conservation and Management Practices

Vegetative

1. Use one- or two-year crop rotation.
2. Use three- or four-year crop rotation.
3. Use five-year or longer crop rotation.
4. Put into permanent pasture.
5. Put into permanent woodland.
6. Do not burn crop residue.
7. Use strip cropping.
8. Use crop residue management.
9. Establish recommended grasses and/or legumes.
10. Protect from burning.
11. Control grazing.
12. Control noxious plants.
13. Control brush or trees.
14. Use only for wildlife or recreation area.
15. Establish grass waterways.
16. Use green manure.
17. Use summer fallow.
18. . . . .

Mechanical

1. Terrace and farm on the contour.
2. Use contour strip cropping.
3. Maintain terraces.
4. Construct diversion terraces.
5. Install a drainage system.
6. Level the land.
7. Clear rocks, brush, and/or trees.
8. Irrigate.
9. Control gullies.
10. Maintain trash cover.
11. . . . .

Fertilizer and Soil Amendments

Lime
Gypsum
Sulfur
Manure
Nitrogen
Phosphorus
Potash
Complete fertilizer
Minor elements
. . . .

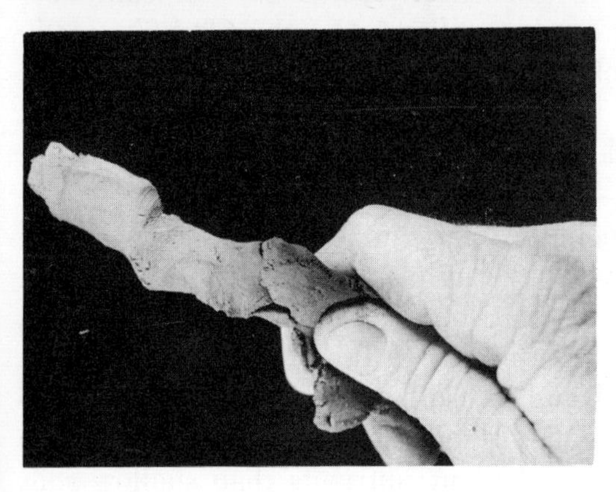

Figure 21-2: Determining Soil Texture. In the field, soil texture is often tested by pressing moist soil into a "leaf" or ribbon. Coarse, sandy soils will form no leaf. Medium-textured soils form a ribbon that breaks easily. Fine-textured, clayey soils make a ribbon longer than one-half inch.

Determining Land Capability Classes

The first step of land judging, covered in Part One of the foregoing table, is the determination of the *land capability class* of a field or a mapping unit. Land capability classes or units are particular types of farm lands, classified according to a formal study of such characteristics as soil texture, depth, slope, and so on. Each of the eight recognized land capability classes has its specific set of soil characteristics. Each is suitable to specific farm uses, and each requires specific soil management procedures. Land capability classes are determined by a study of the following physical features of the land and soil.

Texture of Surface Soil and Subsoil. *Soil texture* may be described as those properties of soils determined by the size of their particles—in other words, by the proportions of *sand*, *silt*, and *clay*. These portions vary in different soils and even in different horizons of the same soil. Soil scientists use several textural classes, such as sand, loamy sand, sandy loam, loam, silt loam, silt, sandy clay loam, clay loam, silty clay loam, sandy clay, silty clay, and clay.

Coarse-textured soils are those that feel gritty. Their dry clods break easily. By the scientist's classification, they include sand, loamy sand, and sandy loam. Coarse-textured soils dry out sooner than fine-textured soils.

Medium-textured soils, when wet, feel floury or loamy between the fingers. They can be rubbed into a short "leaf," usually less than one-fourth inch in length. Dry clods can be less easily broken than those of coarse-textured soils. Medium-textured soils include loam, silt loam, silt, and sandy clay loam. They are easy to cultivate.

Fine-textured soils include clay loam, silty clay loam, sandy clay, silty clay, and clay. They feel plastic and sticky when wet. They can be rubbed into a "leaf" one-half inch or more in length. Small clods, when dry, are difficult to break.

Depth of Surface Soil and Subsoil. Land appraisers pay considerable attention to depth of soil. Deep soils permit roots to reach more soil moisture and plant nutrients than shallow soils with underlying hardpan, rock, or gravel, which may interfere with root development. A further disadvantage of shallow soils is that leveling for irrigation may be impractical. Measurements used for depth of soil include the following:

Deep	36 inches or more.
Moderately deep	20 to 36 inches.
Shallow	10 to 20 inches.
Very shallow	less than 10 inches.

Color of Surface Soil. Soil color classes are generally (1) dark, (2) medium, and (3) light.

Dark-colored soils include those that are black and dark brown. They are usually high in organic matter and fertility. Not all of such soils are fertile, however, some deriving their hue from minerals. Excessively dark soils are likely to be found in swampy areas, which may need drainage.

Medium-colored soils include those that are dark gray, grayish-brown, and light brown.

Light-colored soils include those that are brownish-gray or light gray. Such soils are frequently low in fertility, though light desert soils may be highly productive when irrigated.

Other soil colors prominent in some areas of the United States are red, dark red, light red, and yellow. These are caused by high iron content. Red soils generally have better drainage than yellow soils.

Soil Structure. Soil structure has to do with the arrangement of soil particles into clusters or aggregates. The principal forms of soil structure are platy, prismatic, columnar, blocky, and granular. Soils without structure are (1) single-grained, such as sand in dunes, or (2) massive, such as claypans and hardpans. The soil particles in massive soils adhere without any regular cleavage.

Stoniness. Only a few areas in the United States are entirely free from stones. Degrees of stoniness are (1) free, (2) slight, (3) moderate, and (4) excessive.

Stone free is the term applied to lands containing no stones.

Slight stoniness is the term applied to lands with some stones, but not enough to hinder cultivation seriously.

Moderate stoniness is the term applied to lands that contain enough large stones to make tillage and harvesting difficult.

Excessive stoniness is the term applied to those lands that contain so many large stones as to prohibit cultivation. Such lands are better suited for pasture and trees.

Ease of Cultivation. *Soil tilth* is a term used to classify the physical condition of surface soil in such degrees of consistency as (1) loose, (2) friable, and (3) firm.

Figure 21-3: Friable Soil. Soil like this is easy to cultivate.

Loose soils are those that pour like sand or salt.

Friable soils are those that break and crumble easily. Both loose and friable soils are easy to cultivate.

Firm soils are those that are tough and hard to break even when moist. Land judges classify them as "difficult" or "very difficult" to plow.

Subsoil Permeability. Air and water must be able to penetrate subsoils. Permeability is classed as (1) very slow, (2) slow, (3) moderate, and (4) rapid.

Very slowly permeable subsoils have a dense, heavy clay or claypan composition. They may feel slick or look like putty. If rubbed between the fingers when moist, they will press out into a thin leaf shape.

Slowly permeable subsoils, composed of crumbly clay, have less heavy clay or claypan than very slowly permeable subsoils.

Moderately permeable subsoils are granular. When such subsoils are dry, cracks in the soil profile are perpendicular, permitting easy entrance of air and water, and easy penetration by roots. *These are the most desirable subsoils.*

Rapidly permeable subsoils have considerable coarse sand. They are apt to be droughty because of a low capacity to store soil moisture.

Surface Drainage. Surface drainage is the movement of water across the surface of the soil.

Poor surface drainage is evidenced by standing water or wetness for long periods after rains. Crops will have signs of damage.

Fair surface drainage is evident when water runs but slowly off the land.

Good surface drainage is evidenced by absence of surface water and absence of erosion. The slope and/or soil permeability are sufficient but not excessive.

Excessive surface drainage is associated with slopes of 3 per cent or more. Such fields will exhibit evidences of soil erosion and dryness.

Internal Drainage. Internal drainage is the vertical movement of water down through the soil. A permeable subsoil aids internal drainage, which is particularly necessary in humid and irrigated areas.

Excessive internal drainage is associated with subsoils of sand or loam.

Good internal drainage is indicated by a subsoil that has a uniform color of red, reddish-brown, brown, or yellow.

Fair internal drainage is suggested by a subsoil heavily mottled yellow, or with gray to blue-gray spots.

Poor internal drainage is indicated by a gray to blue-gray subsoil.

Per Cent of Slope. An important consideration in land capability classification is the surface slope, which, in conjunction with such factors as permeability, determines crop and machinery selection, and soil and water conservation measures. As slopes get steeper, surface water flows faster and causes more soil erosion. If the percentage of slope is doubled, soil losses increase 2.5 times. Soils that absorb water readily are less subject to erosion.

Slopes may be classified as (1) nearly level, (2) gently sloping, (3) moderately sloping, (4) strongly sloping, and (5) very steep.

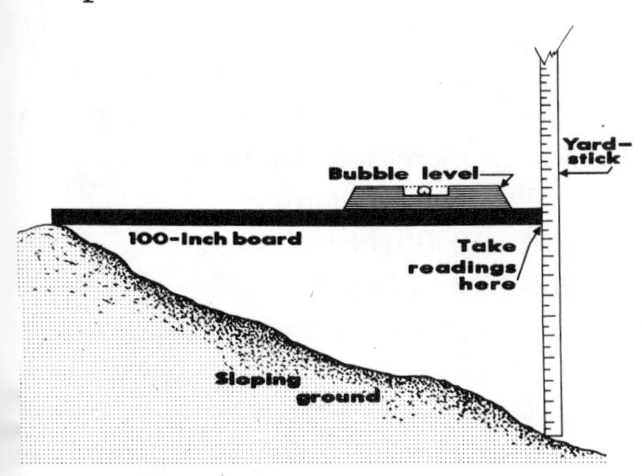

Figure 21-4: Determining Per Cent of Slope. The yardstick shows the vertical distance between the bottom of the leveled, 100-inch board and the ground. This vertical distance in inches is equal to the slope of the ground in per cent.

Slope is measured by the number of feet of rise in each 100 feet of horizontal distance. The percentage of slope can be determined without the use of a land level by means of a 100-inch cord or board. A carpenter's level is used in the middle of this. At the downslope end, the vertical distance to the ground is measured with a yardstick. The vertical distance in inches is equal to the slope of ground in per cent, since the string or board is 100 inches long.

While slope *terms* are similar over the United States, the *degree* of slope may vary for each slope class in different parts of the country. The figures on degree of slope given below are for an area in the Central states.

Nearly level land has less than one foot of fall in each 100 feet. Water will stay on such land if the soil is not absorbent. Some crop damage from poor drainage may result.

Gently sloping land has one to three feet of fall in each 100 feet. This permits water to move over the land at a slow rate. Good drainage can be expected.

Moderately sloping land has three to five feet of fall in each 100 feet. Excessive rains may cause soil erosion because of runoff.

Strongly sloping land has five to eight feet of fall in each 100 feet. If no conservation measures are taken, even ordinary rains will damage cultivated fields.

Very steep land has above eight feet of fall in each 100 feet. Such lands should be kept in permanent vegetation except in areas of light rainfall and permeable soils, where farmers may do some cropping.

Length of Slope. The longer the slope, the more soil and water are lost after cultivation. If the length of the slope is doubled, the soil loss is generally increased 1.5 times.

Long slopes are more than 1000 feet in length.

Medium slopes are from 300 to 1000 feet in length.

Short slopes are less than 300 feet in length.

Degree of Erosion. Erosion from wind or water is associated with loss of topsoil. Depth of topsoil may not be a good measure of such loss because the original topsoil may not have been deep. In addition, present topsoil, although dark in color, may be a mixture of topsoil and subsoil. A better procedure than measuring topsoil depth is to look for such signs of erosion as rills and gullies. Erosion damage may be classified as (1) none to slight, (2) moderate, and (3) severe.

None to slight is the classification applied to erosion of land where less than 25 per cent of the surface soil has been removed. Little or no erosion would be anticipated on level or nearly level lands. There may be soil deposits washed down from neighboring fields, but no rills or gullies.

Moderate is the classification applied to erosion of land from which 25 to 50 per cent of the topsoil has been removed. The surface may show signs of subsoil, especially as light-colored areas on freshly cultivated fields. Low spots will show deposits of soil washed from higher areas. The land surface may have some rocks. Variations will be evident, with more severe erosion in some areas than in others. Sloping land that has been in cultivated crops usually will be moderately eroded.

Severe is the classification applied to erosion of land from which 50 to 75 per cent of the surface soil has been removed. There are rills and starting gullies. Soil deposits are clearly evident at the bottom of slopes. Stone and gravel are frequently found on the land surface. Differences in crop growth within single fields are easily recognized.

Very severe is the classification applied to erosion of land from which over 75 per cent of the surface soil has been removed and in which there are large, uncrossable gullies.

Factors that influence degree of erosion are per cent and length of slope, tillage practices, amount of cover, soil texture, soil structure, and soil drainage.

Erosion Hazard When Cultivated. Erosion hazard when fields are cultivated is usually considered in terms of (1) none to slight, (2) moderate, and (3) severe. The first terms would be applied to deep, easily worked, nearly level land which is not

Figure 21-5: "Very Severe Erosion." Large gullies uncrossable by ordinary farm machinery are the major characteristic of the degree of erosion classified "very severe" in land judging. What management practices would you recommend to stabilize this land?

Figure 21-6: Land Classified by Capability. Seven of the eight land classes are identified in this single view of Sonoma County, California.

subject to flooding. Moderate erosion hazard would be found on lands at least moderately deep and no more than slightly sloping. Such lands are often used for cultivation, but require such management devices as crop rotations, water control, and special tillage practices. Severe erosion hazard would be found on lands with grades of from 4 to 7 per cent. Soils are shallow and sandy on slopes, and subject to severe flooding in flatlands. Cultivation may be practiced only with intensive conservation measures.

Factors Used in Selecting Land Classes

After considering all of the above topics, the land judge reassesses the field in general terms of (1) texture, (2) permeability, (3) depth, (4) slope, (5) erosion, and (6) drainage, in order to determine its land capability class. Some factors may weigh more than others. For instance, a flat, deep, permeable field of good texture may rate a higher classification, in spite of poor subsurface drainage, than a sloping field with somewhat shallower soil and signs of erosion.

Land Capability Classes

Soil conservation specialists have set up a system of land classification, divided into eight *use* or *capability* classes, to simplify the choice of management practices for maintaining or improving productivity.

Classes I to IV are suitable for cultivation, classes V to VIII for permanent vegetation. This division is based largely on soil erosion hazards.

The degree of slope is the most important limiting factor in cultivation. Erosion hazards also increase with the season and the amount and/or intensity of rainfall and wind. Physical characteristics of the soil that may limit cultivation are (1) wetness, (2) stoniness, (3) shallowness of topsoil, and (4) tightness of subsoil.

Land within a single class may be described as (1) very good, (2) good, (3) moderately good, and (4) fairly good, for purposes of estimating its value in the choice of different crops.

Land use classification may vary in different areas of the United States. Specific recommendations for each land class within a region will further vary in respect to kinds of crops and types of tillage practices.

Land Classes Suited for Cultivation

Class I. The soils of Class I lands are deep, easily worked, nearly level, and productive. The lands are not subject to overflow damage. Possible erosion damage is slight.

Class II. The soils of Class II lands are at least moderately deep and subject to no more than moderate erosion. Occasional overflows may necessitate drainage. The above limitations require appropriate conservation and management practices, such as the following:

1. Soil-conserving crop rotations.
2. Water control devices.
3. Special tillage practices.

Subclasses of Class II land may require further combinations of conservation and management practices.

Class III. Slopes of Class III land may range from 4 to 7 per cent. The soils are shallow, sandy, or wet, but can be used for regular cultivation and cropping with intensive conservation and management practices. These may include crop rotation and contour strip cropping.

Class IV. The slopes on some Class IV lands may run from 8 to 15 per cent, with moderate to severe erosion. Other lands in this class may have more moderate slopes, but soils that are shallow and low in fertility. Either type would be better left in permanent vegetation, such as hay or pasture, but may be cropped occasionally if this is done with the most extreme care.

Figure 21-7: Class I Land. This deep, level, fertile, well-watered, and well-drained field of good tilth is very moderately susceptible only to sheet erosion and wind erosion. What management practices would you recommend to control the slight possibility of damage?

Figure 21-8: Class II Land. This deep, almost level, fertile field is prevented from being classifiable as Class I by a limitation of moisture. What management practices would you recommend to hold rainwater and snow on the soil and to prevent wind erosion?

Figure 21-9: Class III Land. Intensive conservation and management practices are needed for regular cultivation of Class III land. See how many such practices you can identify in this picture.

Figure 21-10: Class IV Land. Only with the most extreme care is occasional cropping of Class IV land possible. Such land is much more profitably left in permanent vegetation.

Land Classes Unsuited for Cultivation

Class V. Class V lands are level or nearly so, but unsuited to cultivation because of wetness or stoniness or both. Suitable for pastures, meadows, and woodlands, Class V lands should be kept in permanent vegetation. Grazing should be limited to prevent the injury of plant cover.

Class VI. Class VI land, with slopes running from 18 to 26 per cent, should be used for grazing and forestry. Grazing should again be limited to protect plant cover. Management may include reseeding and fertilizer treatment.

Class VII. Class VII lands include those that are very steep, eroded, rough, shallow, droughty, or swampy. They are suitable for grazing or forestry only with extreme care. Slopes may range from 26 to 60 per cent. Where rainfall is ample, they should be used for woodland; where rainfall is limited, for pasture.

Class VIII. Class VIII lands are so steep, rough, stony, severely eroded, or wet that they are suitable only for wildlife, recreation, and watershed purposes. They should be planted with trees or grass and left undisturbed thereafter.

Figure 21-11: Class V Land. Class V lands are level, but too wet or too stony for profitable cultivation. Grazing or forestry puts these lands to profitable use.

Figure 21-12: Class VI Land. Reseeding, fertilization, liming, and controlled grazing keep Class VI lands safe from erosion.

Figure 21-13: Class VII Land. Ample rainfall makes it most desirable to use this steep Class VII land for carefully managed tree farming. If rainfall were limited, carefully managed pasture would be preferable.

Recommending Land Treatments

Once a farm has been divided into land capability classes, it is fairly easy to determine which good farming practices are needed to conserve soil and water and to maintain and improve productivity. Land treatment is conveniently divided into three parts, (1) vegetative improvement, (2) mechanical improvement, and (3) fertilization and soil amendment.

Figure 21-14: Studying Soil Samples. Samples of topsoil from four different fields have been brought to one place for convenient comparison during this land judging contest. Visual examination helps determine the need for vegetative improvements such as crop rotation or green manuring.

Vegetative Improvement. Included under vegetative improvements are crop selection, crop rotation, strip cropping, handling of crop residues, pasture and range management, grazing control, weed control, and use of green manures and summer fallow. Grassland farming and tree farming are important aspects of crop selection.

Mechanical Improvement. Mechanical improvement includes terracing, contour plowing, drainage, irrigation, leveling, clearing, trash covers, and gully control.

Fertilization and Soil Amendment. At the same time that the farmer is determining his land capability classes, he should be having his soils tested for fertility, reaction, and alkali conditions. With the results of these tests, he can learn which nutrient elements and which soil amendments, such as lime and gypsum, he needs to apply to his soil.

Soil Acidity. Soil acidity may be a consideration in land judging in Eastern states. The acidity of soil is expressed for land judging purposes in the following *p*H values.

Neutral soils range from 6.6 to 7.5 *p*H value.

Slightly acid soils range from 6.1 to 6.5 *p*H value.

Moderately acid soils range from 5.5 to 6.0 *p*H value.

Strongly acid soils have a *p*H value of less than 5.5.

Some land judging events may call for recommendations on the amount of lime materials to be used for correcting soil acidity.

Alkaline, Salty, and Sodic Soils. Western land judging often includes the classification of alkaline, salty, or sodic soils.

Alkaline soils are merely high in lime.

Salty soils are heavily saturated with a thin deposit of salts. Heavy salt deposits destroy or hinder plant growth. Salty soils may have good structure.

Sodic soils once were salty, but later the salts were removed. These soils characteristically have poor structure.

Values of Land Judging Schools and Competitive Events

Land judging schools and contests in recent years have become as popular as livestock judging. Interest in the study of soils has been increased among both youths and adults through these uses of land judging. Some of the values of land judging are as follows:

1. Groups can visit farms where soil conservation and management are being practiced.
2. Groups can come in contact with soil conservation specialists and program leaders.
3. Learning-by-doing is emphasized through actual field and laboratory experience.
4. Land judging schools and field trips provide an incentive for a study of soils.
5. Land judging emphasizes both (1) study of land capability and (2) good management practices.
6. A large number of people are introduced to the need for good soil management.
7. Land judging provides training for farmers in classifying individual fields according to soil characteristics, to determine proper management practices.
8. The principles of land judging are useful to an individual buying a farm or a piece of land.
9. Experience in land judging aids the student in evaluating land for taxation or real estate management.

(N-1)

LAND JUDGING SCORE CARD

Field No.__________

EXTENSION DIVISION
Oklahoma State University

1. (Name or No.) ....
2. (Address) ....
3. (County) ....
4. Soil Conservation District..
5. ....

LAND CLASS FACTORS—PART ONE
Indicate your answer by an X in the square

SURFACE TEXTURE
- Coarse .... [ ]
- Medium .... [ ]
- Fine .... [ ]

MOVEMENT OF AIR AND WATER IN THE SUBSOIL (permeability)
- Very slow .... [ ]
- Slow .... [ ]
- Moderate .... [ ]
- Rapid .... [ ]

DEPTH OF SURFACE SOIL AND SUBSOIL
- Deep .... [ ]
- Moderately Deep .... [ ]
- Shallow .... [ ]
- Very Shallow .... [ ]

SLOPE
- Nearly level .... [ ]
- Gently Sloping .... [ ]
- Moderately Sloping .... [ ]
- Strongly Sloping .... [ ]
- Steep .... [ ]
- Very Steep .... [ ]

EROSION—WIND AND WATER
- None to slight........................ [ ]
- Moderate .... [ ]
- Severe .... [ ]
- Very Severe .... [ ]

SURFACE DRAINAGE
- Poor .... [ ]
- Fair .... [ ]
- Good .... [ ]
- Excessive .... [ ]

INDICATE THE MAJOR FACTORS
Considered in selecting the land class and treatments
- Texture .... [ ]
- Permeability .... [ ]
- Depth .... [ ]
- Slope .... [ ]
- Erosion .... [ ]
- Drainage .... [ ]

LAND CAPABILITY CLASS

I II III IV V VI VII VIII

(Circle one of the above)

SCORE—PART ONE__________
(POSSIBLE POINTS—30)

RECOMMENDED LAND TREATMENTS — PART TWO
(See reverse side for instructions)

[ ]
[ ]
[ ]
[ ]
[ ]
[ ]
[ ]
[ ]
[ ]
[ ]
[ ]
[ ]
[ ]
[ ]
[ ]
[ ]
[ ]
[ ]
[ ]
[ ]

SCORE—PART TWO __________
(POSSIBLE POINTS—30)

SCORE—PART ONE__________
SCORE—PART TWO__________
TOTAL SCORE__________

Figure 21-15: Land Judging Score Card. State agricultural colleges will usually supply sample land judging score cards.

INSTRUCTIONS FOR PART TWO

This is for guidance of participants in selection of proper land treatments needed for different land classes. Select from these the practices needed to conserve soil and water and maintain or improve productivity, and record it by number in the proper square on the opposite side—Part Two. The blank lines may be used for writing in the additional practice when needed.

VEGETATIVE

1. Use soil conserving and improving crops every 4th or 5th year.
2. Use soil conserving and improving crops every 3rd or 4th year.
3. Use soil conserving and improving crops every 2nd year.
4. Use soil conserving and improving crops every year.
5. Do not burn crop residue.
6. Strip cropping.
7. Residue management.
8. Establish recommended grass and/or legumes.
9. Improve by addition of recommended grasses and/or legumes.
10. Proper pasture or range management.
11. Protect from burning.
12. Control grazing.
13. Control noxious plants.
14. Control brush or trees.
15. Plant recommended trees.
16. Harvest trees selectively.
17. Use only for wildlife or recreation area.
18. ....
19. ....
20. ....

MECHANICAL

21. Terrace and farm on contour.
22. Farm on contour.
23. Maintain terraces.
24. Construct diversion terraces.
25. Install drainage system.
26. Control gullies.
27. ....
28. ....
29. ....

FERTILIZER & SOIL AMENDMENTS

30. Lime.
31. Gypsum.
32. Manure or Compost.
33. Phosphate.
34. Potash.
35. Nitrogen.
36. Phosphate and Nitrogen.
37. Phosphate and Potash.
38. Nitrogen and Potash.
39. Phosphate, Nitrogen & Potash.
40. ....
41. ....
42. ....

igure 21-16: Back of Sample Card. What changes would you nake on both sides of this card to use it in your area?

Summary

Land judging is recommended for planning and coordinating soil improvement and management on each farm.

Land judging begins with a determination of the *land capability class* of each *mapping unit* or field. A mapping unit or field is a farm area having approximately the same characteristics and requiring the same crop selection and management practices. Factors considered in classification of mapping units or fields into land capability classes are surface soil texture, subsoil texture, depth of surface soil and subsoil, topsoil color, soil structure, stoniness, ease of cultivation, subsoil permeability (movement of air and water in the subsoil), surface drainage, internal or subsurface drainage, per cent of slope, degree of erosion, and erosion hazard when cultivated. These considerations are then evaluated as a whole, with special attention to texture, permeability, depth, slope, erosion, and drainage. The land judge then decides on the land capability class of the mapping unit.

Classes I to IV are suitable for cultivation. Class I lands are deep, easily worked, nearly level, productive, well drained, and uneroded. Class II lands are moderately deep and subject to no more than moderate erosion. They may need drainage and conservation or management measures such as rotations and special tillage practices. Class III lands have slopes of from 4 to 7 per cent. Soils are shallow, sandy, or wet, and require intensive conservation and management practices for cultivation. Class IV lands have slopes of from 8 to 15 per cent and moderate to severe erosion. Other lands in the same class may be more level, but have shallow, infertile soil. Either type is more suited to permanent vegetation, but may be cropped with extreme care.

Classes V to VIII are unsuited to cultivation. Class V lands are level but unsuited to cultivation because of stoniness or wetness. They are suitable for pastures, meadows, and woodlands but must be protected from overgrazing. Class VI land has slopes of from 18 to 26 per cent and should be used for limited grazing or for forestry. Class VII lands are very steep, eroded, rough, shallow, droughty, or swampy, and are suitable for grazing or forestry only with extreme care. Class VIII lands are so steep, rough, stony, severely eroded, or wet that they are suitable only for wildlife, recreation, and watershed purposes.

Once the farm has been divided into land capability classes, good farming practices are determined for each mapping unit. These include vegetative improvements such as crop rotation, mechanical improvements such as drainage, and fertility and soil amendment such as adding lime or gypsum.

Land judging schools and contests are popular and valuable.

Study Questions

1. What is the history of land judging schools and contests?
2. Why have land judging?
3. What are the two major parts of land judging score cards?
4. With what is the first part of land judging concerned?
5. What are the characteristics of soil texture?
6. How is soil depth classified?
7. How is soil tilth or ease of cultivation classified?
8. How is soil structure described?
9. How is land stoniness classified?
10. How are land slope and length of slope classified?
11. How is soil permeability classified?
12. What kinds of soil drainage are there? Give the subdivisions.
13. Describe the degrees of soil erosion.
14. Classify degrees of acidity and alkalinity.
15. Describe the eight land capability classes.
16. Give examples of vegetative improvements.
17. Give examples of mechanical improvements.

Class Activities

Land judging provides an excellent summary and review of soil science, conservation, and management. Land judging helps create interest in a further study of soil science and approved practices of soil management. For these reasons, it is strongly recommended that the final class activity of this course be a land judging contest.

Suggestions useful to land judging studies and contests are:

1. Score cards are divided into two major parts, namely: (1) physical features of land and soil, and (2) recommended conservation and management practices.
2. Sample score cards should be secured from the state agricultural college and adapted to local conditions.
3. Fields need to be selected for judging. These are available on farms enrolled in the Soil Conservation Program. Information on such fields is contained in the farm's conservation plan.
4. Assistance can be secured from farmers and personnel of the SCS and Agricultural Extension Service.
5. Usually four fields are used in a contest. A single field could be judged as a class or group exercise.
6. Fields should be selected that are typical of the community and a particular land class if the project is a first experience for the group.
7. Permission should be secured from the landowner for use of the land.
8. Care should be taken to avoid damage to the land and to the property. Pits should be filled before leaving, to avoid hazards.
9. Fields should be numbered in advance. A wooden stake with a number is desirable for identifying fields.
10. A posthole digger is a convenient tool for digging pits. Power machinery is timesaving. Shovels can be used to shape the pits.
11. Relatively small pits or holes will prove satisfactory for a small class or training session. In large contests, pits should be three feet deep and wide enough to permit contestants to view or walk into them easily. If sloped on one side, the pit will better enable contestants to view the soil profile.
12. For judging, soil from the pits should be placed in piles or sample boxes and properly labeled as (1) topsoil, (2) subsoil, and (3) parent material, if included.
13. Information should be available to contestants on soil tests for fertility and reaction.
14. Cold weather and snow are unfavorable for land judging programs. A modified type of land judging program can be conducted under cover with the use of soil samples, pictures, charts, and other visual aids.
15. A total of 60 points is used on most land judging score cards, thirty points for each half. On the master score card, judges assign the point values to be given.

GLOSSARY *

"A" horizon—The topmost layer of a mineral soil; topsoil.

"ABC" soil—A soil with a complete profile, including an "A," a "B," and a "C" horizon.

Acid soil—Generally, a soil that is acid throughout most or all of the parts that plant roots occupy; but the term is commonly applied only to the plow layer or to some other specific layer or horizon. Practically, this means a soil more acid than *p*H 6.6; precisely, a soil with a *p*H value less than 7.0; chemically, a soil which in solution has a preponderance of hydrogen over hydroxyl ions.

Actinomycetes—A group of soil microorganisms which produce an extensive, thread-like network. They resemble the soil molds in some respects but bacteria in size.

Aeration, soil—The exchange of air in soil with air from the atmosphere. The composition of the air in a well-aerated soil is similar to that in the atmosphere; in a poorly aerated soil, the air in the soil is considerably higher in carbon dioxide and lower in oxygen.

Aggregate—Many fine soil particles held in a single mass or cluster, such as a clod, crumb, block, or prism. Many properties of the aggregate differ from those of an equal mass of unaggregated soil.

Alkali soil—Generally, a highly alkaline soil; specifically, a soil with so high a degree of alkalinity (*p*H 8.5 or higher), or so high a percentage of exchangeable sodium (15 per cent or higher), or both, that the growth of most plants is reduced.

Alkaline soil—Generally, a soil that is alkaline throughout most or all of the parts occupied by plant roots, although the term is commonly applied to only a specific layer or horizon of a soil; precisely, any soil horizon having a *p*H value greater than 7.0; practically, a soil having a *p*H above 7.3.

Alluvial soils—Soils resulting from transported and relatively recently deposited material (alluvium) with little or no further modification by soil-forming processes. Soils formed from alluvium but with well-developed profiles are grouped with other soils having the same kinds of profiles, not with the alluvial soils.

---

* Glossary items are reproduced from *The 1957 United States Department of Agriculture Yearbook on Soils,* with minor modifications.

Alluvium—Sand, mud, and other sediments deposited by streams.

Amendment—Any material, such as lime, gypsum, sawdust, or synthetic conditioners, that is worked into the soil to make it more productive. Strictly, a fertilizer is also an amendment, but the term is used most commonly for materials other than fertilizer.

Ammonia—A colorless gas composed of one atom of nitrogen and three atoms of hydrogen. Ammonia liquefied under pressure is used as a fertilizer.

Anaerobic—Living or functioning in the absence of air or free oxygen.

Anhydrous—Dry, or without water. Anhydrous ammonia is water free, in contrast to the water solution of ammonia commonly known as "household ammonia."

Anion—An ion carrying a negative charge of electricity.

Aqua ammonia—A water solution of ammonia.

Arid climate—A very dry climate, like that of desert or semidesert regions where there is only enough water for widely spaced desert plants. The limits of precipitation vary widely according to temperature, with an upper limit for cool regions of less than ten inches and for tropical regions of as much as 20 inches.

Arid region—Areas where the potential water losses by evaporation and transpiration are greater than the amount of water supplied by precipitation. In the United States this area is broadly considered to be the dry parts of the 17 Western states.

Ash—The nonvolatile residue resulting from the complete burning of organic matter. It is commonly composed of oxides of such elements as silicon, aluminum, iron, calcium, magnesium, and potassium.

Assimilation—Conversion of substances, taken in from the outside, into living tissue of plants or animals.

Available nutrient in soils—That part of the supply of a nutrient in the soil that can be taken up by plants at rates and in amounts significant to growth.

Available water in soils—The part of the water in the soil that can be taken up by plants at rates significant to their growth; usable or obtainable water.

Azonal soils—A general group of soils having little or no soil profile development. Most of them are young. In the United States, alluvial soils, lithosols, and regosols are included in the azonal group.

"B" horizon—A soil horizon, usually beneath an "A" horizon, or surface soil, in which (1) clay, iron, or aluminum, with accessory organic matter, have accumulated by receiving suspended material from the "A" horizon above it or by clay development in place; (2) the soil has a blocky or prismatic structure; or (3) the soil has some combination of these features. In soils with distinct profiles the "B" horizon is roughly equivalent to the general term "subsoil."

Banding (of fertilizers)—The placement of fertilizers in the soil in continuous narrow ribbons, usually at specific distances from seeds or plants. The fertilizer bands are covered by the soil but are not mixed with it.

Basin irrigation—The application of irrigation water to level areas surrounded by border ridges or levees.

Bedrock—The solid rock underlying soils and other earthy surface formations.

Bench terrace—An embankment, with a steep drop on the downslope side, constructed across sloping soils.

Bog soils—An intrazonal group of soils with mucky or peaty surface soils underlain by peat. Bog soils usually have swamp or marsh vegetation and are commonest in humid regions.

Border irrigation—Irrigation in which the water flows over narrow strips that are nearly level and are separated by parallel, low bordering banks or ridges.

Broad-base terrace—A low embankment, with such gentle slopes that it can be farmed, constructed approximately on the contour across sloping soils. Broad-base terraces are used on pervious soils to reduce runoff and soil erosion.

Brown podzolic soils—A zonal group of soils with thin mats of partly decayed leaves over thin, grayish-brown mixed humus and mineral soil. They lie over yellow or yellowish-brown, acid "B" horizons, slightly richer in clay than the surface soils. These soils develop under deciduous or mixed deciduous and coniferous forests in cool-temperate humid regions, such as parts of New England, New York, and western Washington.

Brown soils—A zonal group of soils having a brown surface horizon that grades below into lighter-colored soil. These soils have an accumulation of calcium carbonate at one to three feet. They develop under short grasses, bunchgrasses, and shrubs in a temperate to cool semiarid climate.

Buffer—A substance in the soil that acts chemically to resist changes in reaction, or *p*H. The "buffering" action is due mainly to clay and very fine organic matter. Thus with the same degree of acidity, more lime is required to neutralize (1) a clayey soil than a sandy soil, (2) a soil rich in organic matter than one low in organic matter, or (3) a sandy loam in Michigan, say, than a sandy loam in central Alabama.

Buffer strips—Established strips of perennial grass or other erosion-resisting vegetation, usually on the contour in cultivated fields.

"C" horizon—The unconsolidated rock material in the lower part of the soil profile from which the upper horizons (or at least a part of the "B" horizon) have developed.

Capillary water—The water retained or raised by the fine pores of soil as a result of capillary forces, including surface tension.

Carbohydrates—Compounds containing carbon, hydrogen, and oxygen. Usually the hydrogen and oxygen occur in the proportion of two to one, as in glucose ($C_6H_{12}O_6$).

Carbon—One of the commonest chemical elements, occurring in lampblack, coal, and coke in varying degrees of purity. Compounds of carbon are the chief constituents of living tissue.

Carbon dioxide—A colorless gas (CO_2) composed of carbon and oxygen, and normally found in small amounts in the air. It is one of the products of the burning (oxidation) of organic matter, or carbon-containing compounds.

Carbon-nitrogen ratio—The ratio of the weight of organic carbon to the weight of total nitrogen in a soil or in an organic material.

Cation—An ion carrying a positive charge of electricity. The common soil cations are calcium, magnesium, sodium, potassium, and hydrogen.

Cellulose—The principal constituent of the cell walls of higher plants.

Chernozem soils—A zonal group of soils having deep and dark-to-nearly-black surface horizons rich in organic matter, which grade into lighter-colored soil below. At 1.5 to four feet, these soils have layers of accumulated calcium carbonate. They develop under tall and mixed grasses in a temperate to cool subhumid climate.

Chestnut soils—A zonal group of soils with dark brown surface horizons, which grade into lighter-colored horizons beneath. They have layers of accumulated calcium carbonate at one to four feet. They are developed under mixed tall and short grasses in a temperate to cool and subhumid to semiarid climate. Chestnut soils occur in regions a little more moist than those having brown soils and a little drier than those having chernozem soils.

Chlorophyll—The constituent responsible for the green color of plants. Chlorophyll is important to photosynthesis in plants, the process by which sugar is manufactured.

Chloroplasts—Small bodies in cells of plants in which the green pigment chlorophyll is concentrated.

Chlorosis—A condition in plants resulting from the failure of chlorophyll to develop, usually because of deficiency of an essential nutrient. Leaves of *chlorotic* plants range from light green through yellow to almost white.

Clay—Mineral soil particles less than 0.002 mm. in diameter; as a soil textural class, soil material that contains 40 per cent or more such particles, less than 45 per cent of sand, and less than 40 per cent of silt.

Clay loam—Soil material that contains 27 to 40 per cent of clay and 20 to 45 per cent of sand.

Claypan—A compact, slowly permeable soil horizon rich in clay and separated more or less abruptly from the overlying soil. Claypans are commonly hard when dry and plastic or stiff when wet.

Clod—A mass of soil, produced by plowing or digging, which usually breaks down with repeated wetting and drying—in contrast to a *ped,* which is a natural soil aggregate.

Colloid, soil—Organic or inorganic matter having very small particle size and a correspondingly large surface area per unit of mass. Most colloidal particles are too small to be seen with the ordinary compound microscope. Soil colloids do not go into true solution as sugar or salt do, but they may be dispersed into a relatively stable suspension and thus be carried in moving water. By treatment with salts and other chemicals, colloids may be flocculated, or aggregated, into small crumbs or granules that settle out of water, though such small crumbs of aggregated colloids still can be moved by rapidly moving water or air. Many soil colloids are really tiny crystals, the minerals of which can be identified with X-rays and other devices.

Colluvium—Mixed deposits of soil material and rock fragments near the bases of steep slopes. The deposits have accumulated through soil creep, slides, and local wash.

Companion crop—A crop grown with another crop, usually a small grain with which alfalfa, clover, or other forage crops are sown. (Formerly such small grain crops were known as "nurse crops," but because the small grain does not truly "nurse" the other crop this older term is being abandoned.)

Complete fertilizer—A fertilizer containing the major elements—(1) nitrogen, (2) phosphorus, and (3) potassium—and possibly trace elements.

Compost—A mass of rotted organic matter made from waste plant residues.

Contour—An imaginary line connecting points of equal elevation on the surface of the soil. A contour terrace is laid out on a sloping soil at right angles to the direction of the slope and level throughout its course. In contour plowing, the plowman keeps to a level line at right angles to the direction of the slope. This usually results in a curving furrow.

Cover crop—A crop grown to cover and protect the soil for a certain part of the year.

Creep, soil—The downward mass movement of sloping soil. The movement is usually slow and irregular and occurs most commonly when the lower soil is nearly saturated with water.

Crumb structure—Very porous granular structure in soils.

Crust—A thin, brittle layer of hard soil that forms on the surface of many soils when they are dry; an exposed hard layer of materials cemented by calcium carbonate, gypsum, or other binding agents.

Most desert crusts are formed by the exposure of such layers, through removal of the upper soil by wind or running water, and their subsequent hardening.

Cytoplasm—The portion of the protoplasm of a cell outside the nucleus.

Damping-off—Sudden wilting and death of seedling plants resulting from attack by microorganisms.

Deep percolation—The downward movement of water beyond the reach of plant roots.

Deep soil—Generally, a soil deeper than 40 inches to rock or other strongly contrasting material.

Desert soils—A zonal group of soils that have light-colored surface soils usually underlain by calcareous material and frequently by hard layers. They are developed under extremely scanty scrub vegetation in warm to cool, arid climates.

Diffusion—The transportation of matter as a consequence of the movement of its constituent particles. The intermingling of two gases or liquids in contact with each other takes place by diffusion.

Dispersion of soil—Breaking up of soil aggregates.

Drainage (a practice)—The removal of excess surface water or excess water within the soil by means of surface or subsurface drains.

Drainage, soil—(1) The rapidity and extent of the removal of water from the soil by runoff and by percolation through the soil; (2) as a condition of the soil, the frequency and duration of periods when the soil is free of saturation.

Drift—Material of any sort deposited by geological processes in one place after having been removed from another. *Glacial drift* includes the materials deposited by glaciers and by the streams and lakes associated with them.

Drought—A period of dryness, especially a long one.

Dry farming—Generally, producing without irrigation crops that require some tillage in subhumid or semiarid regions. The system usually involves periods of fallow during which water from precipitation is absorbed and retained.

Dry weight percentage (of water in soil)—The weight of water expressed as a percentage of the oven-dry weight of soil.

Dune—A mound or ridge of loose sand piled up by the wind. Occasionally, during periods of extreme drought, granulated soil material of fine texture may be piled into low dunes, sometimes called *clay dunes.*

Dust mulch—A loose, dry surface layer of cultivated soil, formerly thought to be effective in reducing the loss of moisture from the underlying soil.

Environment—All external conditions that may act upon an organism or soil to influence its development, including sunlight, temperature, moisture, and other organisms.

Enzymes—Substances produced by living cells which can bring about or speed chemical reaction.

Erosion—The wearing away of the land surface by detachment and transport of soil and rock materials through the action of moving water, wind, or other geological agents.

Essential elements—Those elements necessary for plant growth.

Exchangeable—Pertaining to the ions, in the absorbing complex of the soil, that can be exchanged with other ions. For example, when acid soils are limed, calcium ions are exchangeable for hydrogen ions in the complex; when alkali soils are treated with gypsum, calcium ions are exchangeable for sodium ions that can be leached away.

Exchangeable sodium—Sodium, attached to the surface of soil particles, that can be exchanged with other positively charged ions, such as calcium and magnesium, in the soil solution.

Fallow—Cropland left idle in order to restore productivity, mainly through accumulation of water, nutrients, or both. Summer fallow is a common rotation with cereal grain in regions of limited rainfall. The soil is tilled for at least one growing season to control weeds, to aid decomposition of plant residues, and to encourage the storage of moisture for the succeeding grain crop.

Fertility, soil—The quality of a soil that enables it to provide compounds, in adequate amounts and in proper balance, for the growth of specified plants, when other growth factors such as light, moisture, temperature, and the physical condition of the soil are favorable.

Fertilizer—Any natural or manufactured material added to the soil in order to supply one or more plant nutrients. The term is generally applied to inorganic materials, other than lime or gypsum, sold commercially.

Fertilizer grade—The percentage of plant nutrients in a fertilizer. A 10-20-10 grade contains 10 per cent nitrogen (N), 20 per cent phosphoric acid (phosphorus pentoxide, P_2O_5), and 10 per cent potash (K_2O). This convention is in common use even though nitrogen, phosphorus, and potassium may be present in other forms than these.

Field capacity—The amount of moisture remaining in a soil after the free water has been allowed to drain away into drier soil material beneath; usually expressed as a percentage of the oven-dry weight of soil or another convenient unit.

Field moisture—The water that soil contains under field conditions.

Fine-textured soil—Roughly, clayey soil containing 35 per cent or more of clay.

Flood irrigation—Irrigation by running water over nearly level soil in a shallow flood.

Forage—Unharvested plant material that can be used as feed by domestic animals. Forage may be grazed or cut for hay.

Fungi—Forms of plant life lacking chlorophyll and thus unable to make their own food by photosynthesis.

Granular fertilizer—A fertilizer composed of particles of roughly the same composition, about one-tenth inch in diameter. This kind of fertilizer contrasts with the normally fine or powdery fertilizer.

Granular structure—Soil structure in which the individual grains are grouped into spherical aggregates with indistinct sides. Highly porous granules are commonly called "crumbs." A well-granulated soil has the best structure for most ordinary crop plants.

Gravitational water in soil—The water in the large pores of the soil that drains away under the force of gravity.

Gray-brown podzolic soils—A zonal group of soils having thin organic coverings and thin organic-mineral layers over grayish-brown leached layers that rest upon brown "B" horizons richer in clay than the soil horizon above. These soils have formed under deciduous forests in a moist, temperate climate.

Great soil group—Any one of several broad groups of soil with fundamental characteristics in common. Examples are chernozem, gray-brown podzolic, and podzol.

Ground water—Water that fills all the unblocked pores of underlying material below the water table, which is the upper limit of saturation.

Hardpan—A hardened or cemented soil horizon or layer. The soil material may be sandy or clayey and may be cemented by iron oxide, silica, calcium carbonate, or other substances.

Head—Difference in elevation of water-producing discharge (sometimes used incorrectly for the size of irrigation streams).

Heavy soils—Clayey or fine-textured soils. The term originated from the heavy draught on the horses when plowing.

Horizon, soil—A layer of soil, approximately parallel to the soil surface, with distinct characteristics produced by soil-forming processes.

Humid climate—A climate with enough precipitation to support forest vegetation, although there are exceptions where the plant cover includes no trees, as in the Arctic or high mountains; a climate having a high average relative humidity. The lower limit of precipitation may be as little as 15 inches in cool regions and as much as 60 inches in hot regions.

Humus—The well-decomposed, more or less stable part of the organic matter in mineral soils.

Hydrous—Containing water.

Igneous rock—Rock produced through the cooling of melted mineral matter. When the cooling process is slow, the rock contains fair-sized crystals of the individual minerals, as in granite.

Impervious soil—A soil through which water, air, or roots penetrate slowly or not at all. No soil is absolutely impervious to water and air all the time.

Inorganic—Referring to substances occurring as minerals in nature or obtainable from them by chemical means; these include all matter except the compounds of carbon, but also include carbonates.

Intake rate—The rate, usually expressed in inches per hour, at which rain or irrigation water enters the soil. This rate is controlled partly by surface conditions (infiltration or entry rate) and partly by subsurface conditions (permeability). It also varies with the method of applying water. The same kind of soil has different intake rates under sprinkler irrigation, border irrigation, and furrow irrigation.

Intertilled crop—A crop having or requiring cultivation during growth.

Intrazonal soil—Any one of the great groups of soils having more or less well-developed soil characteristics that reflect a dominating influence of some local factor of relief or of parent material over the normal influences of the climate and the vegetation on the soil-forming processes.

Ion—An electrically charged particle. In soils, an ion is an electrically charged element or combination of elements resulting from the breaking up of an electrolyte in solution. Since most soil solutions are highly dilute, many of the salts exist as ions. For example, all or part of the potassium chloride (muriate of potash) in most soils exists as potassium ions and chloride ions. The positively charged potassium ion is called a *cation* and the negatively charged chloride ion is called an *anion.*

Lacustrine deposits—Materials deposited from lake water. Many nearly level soils have developed from such deposits in lakes that have long since disappeared.

Land—The total natural and cultural environment within which production takes place. *Land* is a broader term than *soil.* In addition to soil, its attributes include other physical conditions such as mineral deposits and water supply; location in relation to centers of commerce, population, and other land; the size of individual tracts or holdings; and existing plant cover, works of improvement, and the like. Some use the term loosely in other senses.

Land-capability classification—A grouping of kinds of soil into special units, subclasses, and classes according to their capability for intensive use and the treatments required for sustained use.

Leaching—The removal of materials in solution by the passage of water through soil.

Level terrace—A broad surface channel or embankment constructed across sloping soil on the contour, as contrasted to a graded terrace, which is built at a slight angle to the contour. A level terrace can be used only on soils that are permeable enough for all of the storm water to soak into the soil so that none breaks over the terrace to cause gullies.

Leveling (of land)—The reshaping or modification of the land surface to a planned grade to provide a more suitable surface for the efficient application of irrigation water and to provide good surface drainage.

Light soil—Sandy or coarse-textured soil.

Lignin—An organic substance that incrusts the cellulose framework of plant cell walls.

Lime—Generally the term *lime,* or *agricultural lime,* is applied to ground limestone (calcium carbonate), hydrated lime (calcium hydroxide), or burned lime (calcium oxide), with or without mixtures of magnesium carbonate, magnesium hydroxide, or magnesium oxide, and materials such as basic slag, used as amendments to reduce the acidity of acid soils. In strict chemical terminology, *lime* refers to calcium oxide (CaO), but by an extension of meaning it is now used for all limestone-derived materials.

Loam—The textural class name for soil having a moderate amount of sand, silt, and clay. Loam soils contain 7 to 27 per cent of clay, 28 to 50 per cent of silt, and less than 52 per cent of sand. (In the old literature, especially English literature, the term "loam" applied to mellow soils rich in organic matter, regardless of the texture. As used in the United States, the term refers only to the relative amounts of sand, silt, and clay; loam soils may or may not be mellow.)

Loamy soil—A general expression for soil of intermediate texture between the coarse-textured or sandy soils, on the one hand, and the fine-textured or clayey soils on the other. Sandy loams, loams, silt loams, and clay loams are regarded as loamy soils.

Loess—Geological deposit of relatively uniform, fine material, mostly silt, presumably transported by wind.

Manure—Generally, the refuse from stables and barnyards, including both animal excrement and straw or other litter. In some other countries the term is used more broadly, to include both farmyard or animal manure and "chemical" manures, for which the term *fertilizer* is nearly always used in the United States.

Marl—An earthy deposit, consisting mainly of calcium carbonate but commonly mixed with clay or other impurities. It is formed chiefly at the margins of fresh-water lakes. It is commonly used for liming acid soils.

Mature soil—Any soil with well-developed horizons in near equilibrium with its present environment.

Mellow soil—A porous, softly granular soil easily worked without becoming compacted.

Metamorphic rock—A rock that has been greatly altered from its previous condition through the combined action of heat and pressure. For example, marble is a metamorphic rock produced from limestone, gneiss is produced from granite, and slate is produced from shale.

Micro-—A prefix meaning "very small," as in *microorganism;* one millionth of something; that which makes use of a microscope, as in *microbiology. Macro-* means "large."

Microorganisms—Forms of life too small to be seen with the unaided eye, or barely discernible.

Mineral soil—A soil composed chiefly of mineral matter, in contrast to an organic soil, which is composed chiefly of organic matter.

Mineralization—The release of mineral matter from organic matter, especially through microbial decomposition.

Minor elements—The essential elements needed in small amounts for plant growth (iron, copper, zinc, manganese, boron, molybdenum, chlorine).

Molecule—A group of atoms bonded together in a characteristic pattern; the smallest portion of a substance retaining chemical identity.

Mottled—Irregularly marked with spots of color. A common cause of mottling in soil horizons is imperfect or impeded drainage, although there are other causes, such as soil development from an unevenly weathered rock. Different kinds of minerals may cause mottling.

Muck—Highly decomposed organic soil material developed from peat. Generally, muck has a higher mineral or ash content than peat and is decomposed to the point that the original plant parts cannot be identified.

Mulch—A natural or artificially applied layer of plant residues or other materials on the surface of the soil. Mulches generally are used to help conserve moisture, control temperature, prevent surface compaction or crusting, reduce runoff and erosion, improve soil structure, and control weeds. Common mulching materials include compost, sawdust, wood chips, and straw. Sometimes paper, fine brush, or small stones are used.

Necrosis—Death associated with discoloration and dehydration of all or parts of plant organs, such as leaves.

Nematodes—Very small worms abundant in many soils and important because many of them attack and destroy plants roots.

Neutral soil—A soil that is neither significantly acid nor alkaline. Strictly, a neutral soil has a *p*H of 7.0; in practice, a neutral soil has a *p*H between 6.6 and 7.3.

Nitrification—The formation of nitrates and nitrites from ammonia (or ammonium compounds), as in soil by microorganisms.

Nitrogen fixation—Generally, the conversion of free nitrogen to nitrogen combined with other elements; specifically, in soils, the assimilation of free nitrogen from the soil air by soil microorganisms and the formation of nitrogen compounds that eventually become available to plants. The nitrogen-fixing microorganisms associated with legumes are called *symbiotic;* those not definitely associated with the higher plants are nonsymbiotic.

Nutrient, plant—Any element taken in by a plant, essential to its growth, and used by it in elaboration of its food and tissue.

Order—The highest category in soil classification. The three orders are zonal soils, intrazonal soils, and azonal soils.

Organic soil—A soil or a soil horizon that consists primarily of organic matter, such as peat, muck, and peaty soil. *Organic* in chemistry refers to the compounds of carbon.

Osmotic—Referring to a type of pressure exerted in living bodies as a result of unequal concentration of salts on both sides of a cell wall or membrane. Water will move from the area having the least salt concentration through the membrane into the area having the highest salt concentration and, therefore, exert additional pressure on that side of the membrane.

Oxidation—A chemical change of an element or compound involving the addition of oxygen or its chemical equivalent.

Oxide—A compound of any element with oxygen alone.

Pan—A layer or soil horizon within a soil that is firmly compacted or is very rich in clay. Examples include hardpans, fragipans, claypans, and traffic pans.

Parent material—The unconsolidated mass of rock material (or peat) from which the soil profile develops.

Peat—Unconsolidated soil material consisting largely of undecomposed or only slightly decomposed organic matter accumulated under conditions of excessive moisture.

Percolation—The downward movement of water through soil.

Permanent pasture—Pasture that occupies the soil for a long time—in contrast to *rotation* pasture, which occupies the soil for only a year or two in a rotation cycle with other crops. As used in the humid parts of the United States, the term *permanent pasture* is equivalent to the European *long ley*.

Permeability, soil—The quality of a soil horizon that enables water or air to move through it.

***p*H**—A numerical designation of acidity and alkalinity, as in soils.

Phase, soil—The subdivision of a soil type or other classificational soil unit having variations in characteristics not significant to the classification of the soil in its natural landscape but significant to the use and management of the soil. Examples of the variations recognized by phases of soil types include differences in slope, stoniness, and thickness because of accelerated erosion.

Photosynthesis—The process of conversion by plants of water and carbon dioxide into carbohydrates under the action of light. Chlorophyll is required for the conversion of the light energy into chemical forms.

Platy soil structure—Structure of soil in which the aggregates have thin vertical axes and long horizontal axes; flat, tabular soil structure.

Plow layer—Surface soil; topsoil.

Podzol—A zonal group of soils having surface organic mats and thin, organic-mineral horizons above gray leached horizons that rest upon illuvial dark-brown horizons developed under coniferous or mixed forests or under heath vegetation in a cool-temperate, moist climate.

Podzolic soil—Soil that has part or all of the characteristics of the podzol soils, especially leached surface soils that are poorer in clay than the "B" horizons beneath.

Pore space—The fraction of the bulk volume or total space within soils that is not occupied by solid particles.

Porosity, soil—The degree to which the soil mass is permeated with pores or cavities.

Prairie soil—A zonal group of soils having dark-colored surface horizons grading through brown soil material to lighter-colored parent material at two to five feet, formed under tall grasses in a temperate, humid climate. The term has a restricted meaning in soil science and does not apply to all soils developed in treeless landscapes.

Prismatic soil structure—Soil structure in which the aggregates are prismlike, with the vertical axes of the aggregates longer than the horizontal axes.

Productivity (of soil)—The present capability of a kind of soil for producing a specified plant or sequence of plants under a defined set of management practices. It is measured in terms of the outputs or harvests in relation to the inputs of production factors for a specific kind of soil under a physically defined system of management.

Profile (soil)—A vertical section of the soil through all its horizons and extending into the parent material.

Protein—Any of a group of nitrogen-containing compounds that yield amino acids on hydrolysis and have high molecular weights. They are essential parts of living matter and are one of the essential food substances of animals.

Protoplasm—The basic, jellylike substance in plant and animal cells; it carries out all their life processes.

Puddled soil—Dense, massive soil artificially compacted when wet and having no regular structure. The condition commonly results from the tillage of a clayey soil when it is wet.

Range (or rangeland)—Land that produces primarily native forage plants suitable for grazing by livestock, including land that has some forest trees.

Reaction, soil—The degree of acidity or alkalinity of a soil mass, expressed in *p*H value.

Red podzolic soils—Formerly, a zonal group of soils having thin organic and organic-mineral horizons over a yellowish-brown leached horizon that rests upon an illuvial red horizon developed under deciduous or mixed deciduous and coniferous forests in a warm to warm-temperature humid climate. These are now placed in the red-yellow podzolic group.

Residual fertilizer—Fertilizer remaining in the soil after one or more cropping seasons.

Root zone—The part of the soil that is invaded by plant roots.

Runoff—The surface flow of water from an area; or, the total volume of surface flow during a specified time.

Saline soil—Soil containing enough soluble salts to impair its productivity for plants, but not containing an excess of exchangeable sodium.

Saline-alkali soil—A soil having a combination of a harmful quantity of salts and either a high degree of alkalinity or a high amount of exchangeable sodium, or both, so distributed in the soil profile that the growth of most crop plants is less than normal.

Salts—The products, other than water, of the reaction of an acid with a base. Salts commonly found in soils break up into cations (sodium, calcium, etc.) and anions (chloride, sulfate, etc.) when dissolved in water.

Sand—Individual rock or mineral fragments having diameters ranging from 0.5 mm. to 2.0 mm. Usually sand grains consist chiefly of quartz, but they may be of any mineral composition. *Sand* is the textural class name of any soil that contains 85 per cent or more of sand and not more than 10 per cent of clay.

Sandy clay—Soil of a textural class containing 35 per cent or more of clay and 45 per cent or more of sand.

Sandy clay loam—Soil of a textural class containing 20 to 35 per cent of clay, less than 28 per cent of silt, and 45 per cent or more of sand.

Sandy loam—Soil of a textural class containing 50 per cent of sand and less than 20 per cent of clay.

Sandy soils—Soils of the sand and loamy sand classes; soil material with more than 70 per cent sand and less than 15 per cent clay.

Sedimentary rock—A rock composed of particles deposited after suspension in water. Chief groups of sedimentary rocks are conglomerates, from gravels; sandstones, from sand; shales, from clay; and limestones, from soft masses of calcium carbonate. There are many intermediate types. Some wind-deposited sands have been consolidated into sandstones.

Seepage—The escape of water through the soil; or the emergence of water from soil along an extensive line of surface, rather than from springs.

Semiarid climate—A climate characteristic of the regions intermediate between the true deserts and subhumid areas.

Series, soil—A group of soils with soil horizons similar in their differentiating characteristics and arrangement in the soil profile, except for the texture of the surface soil, and formed from a particular type of parent material. *Soil series* is an important category in detailed soil classification. Individual series are given proper names from place names near the first recorded occurrence. Thus names such as *Houston, Cecil, Barnes,* and *Miami* are names of soil series that appear on soil maps, and each connotes a unique combination of many soil characteristics.

Silica—An important soil constituent composed of silicon and oxygen; the essential material of mineral quartz.

Silt—(1) Individual mineral particles of soil that range in diameter between the upper size of clay, 0.002 mm., and the lower size of very fine sand, 0.05 mm. (2) Soil of the textural class *silt,* containing 80 per cent or more of silt and less than 12 per cent of clay. (3) Sediments deposited from water in which the individual grains are approximately of the size of silt, although the term is sometimes applied loosely to sediments containing considerable sand and clay.

Silt loam—Soil material having (1) 50 per cent or more of silt and 12 to 27 per cent of clay, or (2) 50 to 80 per cent of silt and less than 12 per cent of clay.

Silty clay—Soil of a textural class containing 40 per cent or more of clay and 40 per cent or more of silt.

Silty clay loam—Soil of a textural class containing 27 to 40 per cent of clay and less than 20 per cent of sand.

Single grain soil—A structureless soil in which each particle exists separately, as in dune sand.

Slip—The downslope movement of a mass of soil under wet or saturated conditions.

Slope—The incline of the surface of a soil. It is usually expressed in percentage of slope, which equals the number of feet of fall per 100 feet of horizontal distance.

Soil—(1) The natural medium for the growth of land plants. (2) A dynamic natural body on the surface of the earth in which plants grow, composed of mineral and organic materials and living forms. (3) The collection of natural bodies occupying parts of the earth's surface that support plants and that have properties due to the integrated effect of climate and living matter acting upon parent material, as conditioned by relief, over periods of time.

A *soil* is an individual three-dimensional body on the surface of the earth unlike the adjoining bodies. (The area of individual soils ranges from less than one-half acre to more than 300 acres.)

A *kind of soil* is the collection of soils that are alike in specified combinations of characteristics. Kinds of soil are given names in the system of soil classification. The terms *the soil* and *soil* are collective terms used for all soils, equivalent to the word *vegetation* for all plants.

Soil characteristic—A feature of a soil that can be seen and/or measured in the field or in the laboratory. These characteristics include soil slope and stoniness as well as the texture, structure, color, and chemical composition of soil horizons.

Soil conservation—The efficient use and stabilization of each area of soil, as needed for use at its optimum level of developed productivity according to the specific patterns of soil and water resources of individual farms, ranches, forests, and other land-management units. The term includes the positive concept of improvement of soils for use as well as their protection and preservation.

Soil survey—The systematic examination of soils in the field and in the laboratory. A soil survey includes description and classification, the mapping of kinds of soil, the interpretation of adaptability for various crops, grasses, and trees, behavior of soil under use or treatment for plant production or for other purposes, and productivity of soil under different management systems.

Solonchak soils—An intrazonal group of soils with relatively high concentrations of soluble salts. These soils are usually light-colored, without characteristic structural form, developed under salt-loving plants, and occurring mostly in a subhumid or semiarid climate. In soil classification, the term applies to a broad group of soils and is only approximately equivalent to the common term "saline soil."

Solonetz soils—An intrazonal group of soils having surface horizons of varying degrees of friability underlain by dark-colored hard soil, ordinarily with columnar structure (prismatic structure with rounded tops). This hard layer is usually highly alkaline. Such soils are developed under grass or shrub vegetation, mostly in subhumid or semiarid climates. The term is used for a broad group of soils that include many so-called "alkali" soils in the Western part of the United States. (Where the hard, clayey layer is overlain with a light-colored leached layer, the soils are called *solodized solonetz.*)

Solum—The upper part of a soil profile, above the parent material, in which the processes of soil formation are active. The solum in mature soils includes the "A" and "B" horizons. Usually the characteristics of the material in these horizons are quite unlike those of the underlying parent material. The living roots and other plant and animal life characteristic of the soil are largely confined to the solum.

Storage capacity—The amount of water that can be stored in the soil for future use by plants and for evaporation.

Strip cropping—The practice of growing crops in a systematic arrangement of strips, or bands. Commonly, cultivated crops and sod crops are alternated in strips to protect the soil and vegetation against running water or wind. The alternate strips are laid out approximately on the contour of soils subject to water erosion, at approximate right angles to the prevailing direction of the wind on soils subject to wind erosion.

Structure, soil—The arrangement of primary soil particles into compound particles or clusters that are separated from adjoining aggregates and have properties unlike those of an equal mass of unaggregated primary soil particles. The principal forms of soil structure are platy, prismatic, columnar (prisms with rounded tops), blocky (angular or subangular), and granular. Structureless soils are (1) single grain—made up of unaggregated individual particles, as in dune sand, or (2) massive—made up of particles adhering together without any regular cleavage, as in many claypans and hardpans. ("Good" or "bad" tilth are terms for the general structural condition of cultivated soils according to particular plants or sequences of plants.)

Stubble mulch—A mulch consisting of the stubble and other crop residues left in and on the surface of the soil as a protective cover during the preparation of a seedbed and during at least part of the growing of the succeeding crop.

Subhumid climate—A climate intermediate between semiarid and humid with sufficient precipitation to support a moderate to heavy growth of short and tall grasses, or shrubs, or of these and widely spaced trees or clumps of trees.

Subirrigation—Irrigation through controlling the water table in order to raise it into the root zone. Water is applied in open ditches or through tile until the water table is raised enough to wet the soil. Some soils along streams are said to be naturally "subirrigated."

Subsoil—The "B" horizon of soils with distinct profiles. In soils with weak profile development, the subsoil can be defined as the soil below the plowed soil (or its equivalent of surface soil), in which roots normally grow. Although a common term, it cannot be defined accurately. It has been carried over from early days when "soil" was defined as the plowed soil and everything under it as the "subsoil."

Subsoiling—The tillage of the soil below the normal plow depth, usually to shatter a hardpan or claypan.

Subsurface tillage—Tillage with a sweep-like plow or blade that does not turn over the surface cover or incorporate it into the lower part of the surface soil.

Surface soil—The soil ordinarily moved in tillage, or its equivalent in uncultivated soil, about five to eight inches in thickness.

Synthesis—Combination of simple molecules to form another substance—for example, the union of carbon dioxide and water to form carbohydrates under the action of light in photosynthesis.

Terrace—An embankment or ridge constructed across sloping soils on the contour or at a slight angle to the contour. The terrace intercepts runoff in order to retard it for infiltration into the soil or to conduct it harmlessly to a prepared outlet.

Textural class—A kind of soil material defined according to the proportions of sand, silt, and clay. The principal textural classes of soil, in order of increasing amounts of silt and clay, are as follows: Sand, loamy sand, sandy loam, loam, silt loam, silt, sandy clay loam, clay loam, silty clay loam, sandy clay, silty clay, and clay. These class names are modified to indicate the size of the sand fraction or the presence of gravel, cobbles, and stones. For example, terms such as *loamy fine sand, very fine sandy loam, gravelly loam, stony clay,* and *cobbly loam* are used on detailed soil maps. These terms apply only to individual soil horizons or to the surface layer of a soil type, as in the name *Miami silt loam.* Commonly, the various horizons of any one kind of soil belong in different soil textural classes.

Texture, soil—The relative proportions of the various size-groups of individual soil grains in a mass of soil. Specifically, the term refers to the proportions of sand, silt, and clay.

Tillage—The operation of implements through the soil to prepare seedbeds and rootbeds.

Tilth, soil—The physical condition of a soil in respect to its fitness for the growth of a specified plant or sequence of plants. Ideal soil tilth is not the same for each kind of crop nor is it uniform for the same kind of crop growing on contrasting kinds of soil.

Topography—The shape of the ground surface, as determined by such major features as hills, mountains, or large plains. "Steep" topography indicates steep slopes or hilly land; "flat" topography indicates flat land with minor undulations and gentle slopes.

Topsoil—(1) A presumably fertile soil or soil material, usually rich in organic matter, used to topdress roadbanks, lawns, and gardens. (2) The plow layer of a soil; surface soil. (3) The original or present dark-colored upper soil, which ranges in depth from a mere fraction of an inch to two or three feet on different kinds of soil. (4) The original or present "A" horizon, varying widely among different kinds of soil. Applied to soils in the field, the term has no precise meaning unless defined as to depth or productivity in relation to a specific kind of soil.

Trace elements—Elements found in plants in only small amounts, including several that are essential to plant growth, others that are essential to animals even though not to plants, and others having no known biological functions.

Transpiration—Loss of water vapor from the leaves and stems of living plants into the atmosphere.

Type, soil—A subgroup or category under the soil series, based on the texture of the surface soil. A soil type is a group of soils having horizons similar in differentiating characteristics and arrangement in the soil profile and developed from a particular type of parent material. The name of a soil type consists of the name of the soil series plus the textural class name of the upper part of the soil, equivalent to the surface soil. Thus *Miami silt loam* is the name of a soil type within the Miami series.

Upland—High ground; ground elevated above the lowlands along rivers or between hills.

Virgin soil—A soil that has not been significantly disturbed from its natural state.

Viscosity—Property of stickiness of liquid or gas due to its cohesive and adhesive characteristics.

Volatilization—The evaporation or changing of a substance from liquid to vapor.

Water requirement (of plants)—Generally, the amount of water required by plants for satisfactory growth during the season; more strictly, the number of units of water required by a plant during the growing season in relation to the number of units of dry matter produced. The water requirement varies with climatic conditions, soil moisture, and soil characteristics. Factors unfavorable to plant growth, such as low fertility, disease, and drought, increase the water requirement.

Water table—The upper limit of the part of the soil or underlying rock material that is wholly saturated with water. In some places an upper, or "perched," water table may be separated from a lower one by a dry zone.

Water-holding capacity—The capacity (or ability) of soil to hold water; "field" capacity is the amount of water held against gravity. The water-holding capacity of sandy soils is usually low, while that of clayey soils is high. This capacity is often expressed in inches of water per foot of soil depth.

Waterlogged—A condition of soil in which both large and small pores are filled with water. The soil may be intermittently waterlogged because of a fluctuating water table or waterlogged for short periods after rain.

Watershed—In the United States, the total area contributing water to a stream at a given point; in some other countries, the topographic boundary separating one drainage basin from another. American synonyms are *drainage basin* or *catchment basin.*

Weathering—The physical and chemical disintegration and decomposition of rocks and minerals.

Wilting point (or permanent wilting point)—The moisture content of soil, on an oven-dry basis, at which plants wilt and fail to recover their turgidity when placed in a dark humid atmosphere.

Yellow podzolic soils—Formerly, a zonal group of soils having thin organic and organic-mineral layers over grayish-yellow leached horizons that rest on yellow "B" horizons, developed under coniferous or mixed coniferous and deciduous forests in a warm-temperate to warm, moist climate. These soils are now combined into the red-yellow podzolic group.

INDEX

N

O

P

Q

R

S

Y

Z

6 7 8 9 10